Designing the Purposeful World

T0258748

In September 2015, at the United Nations, world leaders agreed on seventeen Sustainable Development Goals or SDGs. This book extrapolates the SDGs into the idea of a purposeful world. In this context, the purpose for humanity is to thrive sustainably alongside other life forms and to consciously celebrate the process. The SDGs serve as a powerful vision, time-stamped at the 2030 time horizon, not just for world leaders but for us all. However, faced with the challenges of implementing the SDGs, we (including business leaders, government leaders and anyone wishing to make a difference) can feel overwhelmed. Wilson takes the reader on a journey of thought and invites them to work out their personal role in sustainability as well as their collaborative role alongside others in their communities and organisations. Written in a very accessible style, the book celebrates some of the many achievements made by ordinary people as a catalyst for hope, sets out a number of achievable goals and provides exercises to enable the reader to adopt practices that help to make a difference. It is the perfect book to help turn the SDGs into action at every level – governmental, organisational and personal.

Clive Wilson chairs the United Nations Association (Harrogate) as an advocate for the UN's Sustainable Development Goals. He is an experienced consultant and is committed to both organisational sustainability as well as improving the effectiveness of social entrepreneurs in the developing world.

"Clive Wilson is as motivating in print as he is in person. In his latest book he combines a wealth of knowledge, his own practical life experiences and those of others, to convey purpose and value. All this in a world where action is essential but the means is often lacking. Written with clarity and free from jargon, Clive accompanies the reader along a learning journey, as he builds progressive stages of the book. The reader receives continual encouragement to take a hands-on approach towards achieving a purposeful world by applying our own thoughts and experiences throughout. The result is a guide with resounding value, rather than merely food for thought, as is often found in other reflective, less interactive books.

With humanity on the brink of destroying life on our planet, this volume turns the situation around. Instead of painting a picture of gloom and doom, Clive embraces our challenge as an opportunity. Using the framework of the 17 Sustainable Development Goals (SDGs), we are shown a practical pathway for each of us to achieve success. Clive develops the themes from *Designing the Purposeful Organization*, adeptly ramping these up to the present-day global scenario. Accordingly, the reader has much to gain whether or not they have read his previous insights. If you are looking for an accessible, real-world methodology for tackling the global environmental and social issues facing us, you are bound to find unpretentious inspiration in this book."

Michelle Marks, Coral Mountain

"In writing *Designing the Purposeful World*, Clive hasn't simply published a book, he has created a portal to an adventure that will change your world and that of future generations. His work will take you on pathways that twist and turn, stopping off to hear from inspiring individuals with equally impressive stories along the way. The book will challenge your way of thinking, make you question the way the world works and almost certainly support you in making changes that will improve your way of life and support a more environmentally sustainable world."

Nathan Atkinson, Founder of Fuel for School
and Trustee of The Real Junk Food Project

"A powerful employee engagement opportunity – to help make a positive difference, no matter how small. This book is an excellent follow-up to *Designing the Purposeful Organization: how to inspire business performance beyond boundaries*. Overall, an insightful and uplifting read for everyone, written in a practical and down-to-earth conversational style. This book raises awareness of the UN's Sustainable Development Goals and encourages the reader to reflect on the associated meaningful impacts to everyday life, whether this be on a personal, community or work basis."

Melanie Cheung, HR Professional, NHS (UK)

Designing the Purposeful World

The Sustainable Development Goals as a Blueprint for Humanity

Clive Wilson

Routledge
Taylor & Francis Group

LONDON AND NEW YORK

First published 2018
by Routledge
2 Park Square, Milton Park, Abingdon, Oxon OX14 4RN

and by Routledge
711 Third Avenue, New York, NY 10017

Routledge is an imprint of the Taylor & Francis Group, an informa business

© 2018 Clive Wilson

The right of Clive Wilson to be identified as author of this work has been asserted by him in accordance with sections 77 and 78 of the Copyright, Designs and Patents Act 1988.

All rights reserved. No part of this book may be reprinted or reproduced or utilised in any form or by any electronic, mechanical, or other means, now known or hereafter invented, including photocopying and recording, or in any information storage or retrieval system, without permission in writing from the publishers.

Trademark notice: Product or corporate names may be trademarks or registered trademarks, and are used only for identification and explanation without intent to infringe.

British Library Cataloguing-in-Publication Data
A catalogue record for this book is available from the British Library

Library of Congress Cataloging-in-Publication Data
Names: Wilson, Clive, 1955– author.
Title: Designing a purposeful world : the sustainable development
    goals as a blueprint for humanity / Clive Wilson.
Description: Edition. | New York : Routledge, 2018. |
    Includes index.
Identifiers: LCCN 2017050634 | ISBN 9780815381334
    (hardback) | ISBN 9780815381327 (pbk.)
Subjects: LCSH: Sustainable development. | Economic
    development—Environmental aspects. | Social change.
Classification: LCC HC79.E5 W5295 2018 | DDC 338.9/27—dc23
LC record available at https://lccn.loc.gov/2017050634

ISBN: 978-0-8153-8133-4 (hbk)
ISBN: 978-0-8153-8132-7 (pbk)
ISBN: 978-1-351-21068-3 (ebk)

Typeset in Garamond
by Apex CoVantage, LLC

Contents

Figures

Foreword

Who are you? Why are you here? What is your purpose? Most people agree that these are important questions which we should all reflect upon because they strongly influence our motivations. But I believe *the* most important questions, in terms of our collective future, are *"How do I identify myself?"* and *"What do I call my home?"* Because who and what you identify with is what you care for.

Do you identify with your clan, with your tribe, with your state, with your nation, with the people of your religion, with the people of your race, or do you identify with humanity? Whichever group you identify with not only determines who you care about but also determines your worldview.

The worldview that is now emerging is Humanity Awareness. This is the first worldview that is truly systemic and integral. It takes decision making to a new level; it looks at the big picture considering the whole rather than the parts; it focuses on the needs of our global society.

Humanity Awareness unites all peoples of the planet – people of different ethnicity, nationality, race and religion – in an overarching cosmology that unifies the belief systems of spirituality, science and psychology in an energetic framework of understanding, where every human being is recognised as an individuated aspect of the universal energy field – as an energetic soul experiencing three-dimensional reality in a physical body. In other words, in Humanity Awareness everyone on the planet shares the same sense of identity and shares the same values. And, we all share the same home – the planet we call earth.

This is the worldview that Clive Wilson lives from, the worldview he embraces wholeheartedly in everything he does. In my view, that is why he has written this book. He is living his purpose. He invites you to live your purpose by participating in the adventure of making the world a better place for everyone. This book is an invitation to us all to live in Humanity Awareness. The vehicle Clive has chosen to unite our efforts is the United Nations' Sustainable Development Goals.

Throughout human history, individuals and societies have always focused their energies on a central key idea that could, if sufficient energy and effort were devoted to it, help them move towards a more idealised future. The central key idea, at this point in our human history, is to reduce the inequalities that our economic system has produced by helping the world's poorest people achieve minimum standards of income, health care and education, and at the same time promote a sustainability agenda which enables us to safeguard our home – the planet. This central key idea has become known as the "development agenda". The vision and goals for the current development agenda are expressed in the SDGs which were set out by world leaders at the UN in September 2015. I believe the key obstacle to implementing the UN's SDGs is the lack of Humanity Awareness.

If, as a global society, we really do want to build a sustainable future for everyone, there needs to be a seismic shift in the psychological development of our political leaders: a shift from a focus on "I" to "we"; and a shift in attitude from what's in it for me, to what's best for the common good. It is very clear we will not solve the issues we face as a global society until we experience an evolution of human consciousness. What our world leaders are failing to understand is that there is an evolutionary advantage in being able to expand your consciousness (your sense of identity) to include others – in other words, there is an evolutionary advantage in advancing our collective psychological development.

This is the encouragement Clive Wilson offers in this accessible text, not just to world leaders but to us all.

Richard Barrett, author of *Love, Fear and*
the Destiny of Nations
2 November 2017

Acknowledgements

I would like to express my thanks for the support of some very special people, without whom I might never have published this book:

- My wife Frances, my children and other family members for their encouragement and for putting up with my absence from family time whilst writing this book. Especially when I completely disappeared to Seahouses for days on end!
- My colleagues at Primeast who allowed me to reduce my "firm-time" to embark on a journey somewhat different to our core business of leadership, organisational change and team development. Especially for their encouragement in doing so.
- My clients and other connections in the business world, especially Phil Clothier, Richard Barrett, John Campbell and Christophe Horvath who are enthusiastic fellow travellers of this SDG journey.
- My publisher, especially Rebecca Marsh, who has encouraged me in my writing ever since our first meeting in Harrogate in 2016.
- Our UNA group, members or not, who share their challenges and encourage each other in delivering their part for the SDGs.
- The schools, universities, institutions and businesses who have engaged me to run workshops aimed at raising awareness to the SDGs.
- And last, but not least, you the reader for joining me on this journey.

Chapter 1

Introduction

Why any one of us can make a difference

Right at the outset, I want people to realise that this book is for them, whether they respond at a personal, community or corporate level. The goals apply everywhere and to everyone and even seemingly small personal actions add up. They also begin to set trends that others follow. This chapter is set to inspire and engage.

This is a book of action from everyone

The whole essence of this book is that we can all participate in one of the greatest adventures known to man, that of making the world a better place for everyone to enjoy and pass to future generations. Participation is a key thread that will run through the pages that follow. It is my intention to provide you, the reader, with numerous activities that will take you beyond the pages of the book and into a world of inquiry and of possibility. I trust you will find the activities thought-provoking and inspiring.

Activity 1.1: The world we want

Close your eyes for just a few minutes and take some time to reflect before you begin reading this book. I want you to put yourself into the year 2030 and imagine the world you would like to see. Imagine a world you would be happy to pass to future generations. Without me guiding or prompting you too much, take time to explore the world at this time horizon. Feel free to explore the planet and visit places near and far.

Now, in the space below, make as many notes as you are able of what you saw in your mind's eye.

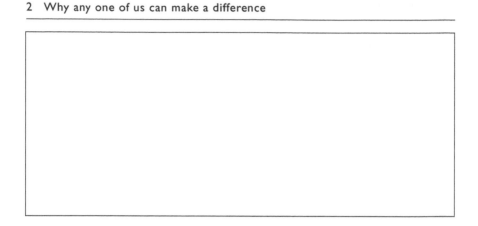

I have tried this exercise many times, with myself, with other people aged from 7 to 70 and with groups of people numbering from five to 500 at workshops and conferences around the world. These have mainly been young people in schools, universities and youth groups but I've also engaged with retirees and people in a variety of businesses and professional institutions.

We are aligned

Curiously enough, what people see is consistent. With very few exceptions, people see a world that is identical to the one world leaders have already signed up to deliver. And the amazing thing is that about 90% of the people I have engaged with had no idea that such a promise had been made. Even when I ask them what significant world event happened in New York in September 2015, most don't know that world leaders signed up to a set of seventeen Sustainable Development Goals (SDGs), with a target date of 2030, that promise to transform our world in a manner that is unprecedented.

The goals are represented in Figure 1.1 below or alternatively, feel free to read the full descriptions of the seventeen goals which you can find on the United Nations Sustainable Development Knowledge Platform (sustainabledevelopment.un.org). This Knowledge Platform is, by the way, a wonderful resource for information relating to the SDGs.

Now, of course, it isn't that we all see exactly the same world when we imagine what things could be like in 2030. But from my experience of engaging with thousands of people, I can confidently affirm that the vast majority of us see something that aligns to some aspects of the SDGs.

The future of our world is written on our hearts

The wonder of this, which I see as nothing short of a miracle, is that we actually don't need to be taught about the goals. It's as if they are already placed on our hearts. However, as I have indeed learnt more about this amazing set of goals, I have found myself becoming more and more inspired.

Figure 1.1 The Sustainable Development Goals
Source: United Nations

Activity 1.2: My personal inspiration

I have a second activity for you now. I want you to scan the seventeen goals one by one, long enough just to get a sense of what they are. You can do this by looking at Figure 1.1, or the more detailed images and descriptions on the UN Sustainable Development Knowledge Platform. But don't spend too long over-thinking them. Just notice how you are drawn to some goals more than others. Make a note of your top goal (or goals) in the space below and ask yourself in what way you are attracted to that goal. Add your thoughts to your notes. Perhaps even make a note of anything you already feel inclined to do. Write that down.

A global commitment

Now let's turn our attention to the world leaders that met in New York in September 2015. The fact that almost all the heads of state across our planet came to together and agreed to transform our world is no small accomplishment. A great deal of commendation must also go to the members of the working party who drafted the goals on their behalf.

I guess they were inspired by the significant accomplishments that the world made in response to the Millennium Development Goals or MDGs that leaders committed to in 2000 and which had a target date of 2015. For example, 43 million more children now go to school; new HIV infections are reduced by 40%; 2 billion more people have clean drinking water; and extreme poverty has been halved.

Activity 1.3: No point going halfway

It's time for another activity if you have access to the Internet. Go to YouTube and search for "no point going half way". This inspiring video is only two and a half minutes long and quickly convinces us that we are on a journey, with more distance to travel. Record your thoughts below.

How do you feel about the commitment our world leaders have made? The reactions of people I meet vary considerably. Most people are pleased, many are inspired. However, some are rightly sceptical, reminded of so many promises that leaders make that get broken.

Look at it this way. Even if you have your doubts that world leaders will deliver on the SDGs, at least the goals have been set. And the goals are supported by measurable targets that will track progress year-on-year. So our leaders can hold each other to account. And we can also hold them to account through the political processes of our world.

We can all play our part

The wonderful thing about the goals is what I said earlier on. The goals are written on our hearts. So, if a few billion, or even a few million people simply follow their hearts, I reckon, we can and will do this together.

As I will demonstrate later in this book, the way to achieve anything in this life is through a sense of purpose. Being conscious of purpose allows a vision to arise and, if we know how, we can create the conditions in which the vision (and hence the purpose) is manifest or allowed to happen.

Because we are all different and inhabit different contexts, we will each have different purposes. That is why we are each drawn to certain SDGs. Even two people who are drawn to the same goal could well be drawn in different ways.

Take for example two men who attended (different) workshops I ran in universities. Both were attracted to SDG 14: life below water. One enjoyed scuba diving in his leisure time and became inspired to make his career cleaning up plastic from the oceans. The other began thinking about fish stocks and how his purchasing choices could make a difference in preserving biodiversity in our seas.

Every action counts

The point is this: every commitment that we feel inclined to make does indeed make a difference. None is too small or insignificant. This is not a competition. This is not something we need to lecture people on. All we have to do is raise awareness of the SDGs and provide space in which people can make new commitments, however small they may seem to be.

None of us has to deliver the SDGs on our own. We just have to listen to our hearts and act accordingly. Every time we take even the smallest, humblest of actions in support of the SDGs, we move the world closer to delivering the greatest vision humankind has ever had.

The power of a ripple

> *I alone cannot change the world,*
> *But I can cast a stone across the waters to create many ripples.*
>
> —Mother Teresa

My experience has also taught me the power of the ripple effect. Actions have reactions, in ourselves and in others. Let me give you a personal example. A few years ago, my son's teacher decided to take up a position running an infant home in Malawi. As a professional speaker, I decided to donate some of my speaking fees to support the home. At that time I had never been to Africa. But this one act of faith set up a chain of events that would take me personally to nine countries in this amazing continent. I have spoken at conferences, run workshops and even helped to establish a consultancy dedicated to supporting the agencies of the United Nations and other organisations who are actively working to deliver the SDGs. Such was the ripple effect working for me.

As an example of the ripple effect working in others, I learned of an attendee at one of my workshops who had been inspired by SDG 4: education. She wrote to tell me that, after much thought and planning, she was heading for Zambia to set up a much needed school there. I wonder what subsequent ripples she will cause to flow from her actions.

Perhaps now is a good time to look back to your notes on the exercises above. Is there anything you need to add to the thoughts and commitments you have noted? Maybe a deadline or the name of someone to consult with?

The power of social media

So, as we have seen, each and every one of us can do something to deliver the SDGs. With the power of social media, we can spread the word to encourage others. Being very aware of the power of the Internet, I have set up a Facebook page called SDGs (the full page title is SDG2030). You're very welcome to join me there. I use it to encourage and celebrate positive action in support of the goals. The number of people following this page is rising exponentially.

I expect that, by the time this book is published we will be a community of over ten thousand people. Feel free to like and share this page to spread the word about the greatest commitment ever made by humanity.

The energy of youth

One of the reasons I have delivered the majority of my workshops in schools, universities and forums for young people is that they have so much energy and enthusiasm to change the world. Of course, they have a significant vested interest, for this is the world they must inhabit!

Young people can use the SDGs as inspiration for their education and career choices. So, this is probably a good time for me to mention an important feature of the goals. They apply to the whole world. In this way I believe they go significantly beyond the Millennium Development Goals which were predominantly aimed at the so-called developing world.

Therefore, becoming a doctor, nurse or sports instructor in Europe is absolutely in line with SDG 3: good health. And being a teacher in Australia is absolutely contributing to SDG 4: education, as, I would suggest, is being a good parent or guardian.

Because engaging young people is so important, one of the first actions I took in 2015 was to begin visiting schools, universities and youth groups to share the SDGs. The ripple effect of this was being invited to facilitate at the Youth Action Summit of December 2015 at the United Nations in New York. The event had been organised by AIESEC International, the world's largest youth-led youth movement comprising around 100,000 young members in their early 20s who seek leadership development in organisations around the world. Just search the Internet for more about the work of AIESEC.

At the Youth Action Summit, I was honoured to share my simple method for engaging young people and I was humbled when AIESEC subsequently made the commitment to "reach every young person everywhere" and inspire them with the SDGs. To aid them in their quest, I shared my slide deck and drafted a set of facilitator notes to help them deliver powerful workshops in just an hour. You too can have these materials by contacting me directly.

Case study 1.1: Young people are determined to change the world

Keen to check on the progress made by the young people at AIESEC, I caught up with Federico Restrepo Sierra who had been one of their enthusiastic leaders for the SDG work. He reminded me of the quote from Margaret Mead below and gave me his version of events during and following the Youth Action Summit:

> *Never doubt that a small group of thoughtful, committed citizens can change the world; indeed, it's the only thing that ever has.*

Few times in history, humanity has come together to dream of a better tomorrow for everyone. One of those times was in 2015 when, in the United Nations Headquarters in New York, the world came together to announce a Global Agenda, seventeen bold goals that are integrated and indivisible, balancing the three dimensions of Sustainable Development: economic, social and environmental.

The seventeen Sustainable Development Goals (SDGs) represent a better world that can only be achieved if we are all united in action and the concept of *Leave no one behind*, generating a call to action to everyone to line behind this ambition.

Inspired by the commitments of all the world leaders, AIESEC decided to play an active and leading role in this new era. If the goals are to be achieved, then young people must play an active role contributing every day for a better today and tomorrow. As the world largest youth-led organisation, we decided to contribute and together with Asian Development Bank (Represented by Chris I. Morris), UN-Habitat (Douglas Ragan, Dana Podmolikova), PVBLIC Foundation (Eliane Sussman, Karolina Piotrowska), Plan International (John Trew), Electrolux (Malin Ekefalk, Carla Silveira) and Primeast (Clive Wilson, John Campbell); AIESEC (Federico Restrepo Sierra, Victoria de Mello) created Youth for Global Goals, a global initiative to mobilise every young person to achieve the Sustainable Development Goals. The knowledge and experience from the different people that decided to step up will be felt in 2030 when we together celebrate the achievement of this great agenda. Our main driver is and will continue to be young people and how we can use this great generation to eradicate poverty and reverse climate change, with youth no longer being a target but an active player in global matters, *Youth for Global Goals* is an initiative for young people.

Youth for Global Goals has three dimensions:

1. **Awareness:** a young person is aware when they know that:

    a. The SDGs are due by 2030
    b. The SDGs are for a better world
    c. The SDGs need collaboration to be achieved

2. **Understanding:** a young person understands when they know:

    a. How to generate impact projects to contribute to at least one SDG
    b. The targets of at least one SDG
    c. AIESEC host YouthSpeak Forums all around the world to promote this knowledge among young people

3. **Action:** a young person is taking action when they contribute to the achievement of at least one SDG. AIESEC runs over five thousand projects every year for young people to join and contribute to a better world.

During 2016, AIESEC was able to reach over 12 million young people (on Awareness), over ninety thousand young people (on Understanding through attendance to over 250 YouthSpeak Forums) and over thirty thousand young people joining our projects.

Analysis versus instinct

Some people will want to explore the goals in their detail and select specific targets to aim for. Others will have no need for such specifics and will simply be inspired to follow their hearts. You see, I believe the vision presented by the SDGs is far bigger than we think it is.

Take SDG 1: no poverty. There are some very specific targets that support this goal about what defines extreme poverty. It is right that governments take action to lift people above such limits. But poverty exists at monetary levels way above a few dollars a day of income. There are people on good salaries who can't make ends meet because illness or redundancy has affected their partner's earning capacity. They may face home repossession or worse. In my humble view, anyone working to support people in such difficulty is surely a SDG 1 crusader.

Think positive

One slight difficulty I have with some of the SDGs, especially SDG 1 is the title being one of negativity. Even though the SDG image clearly says "no poverty", I have learned through studying neurolinguistic programming (NLP) that our brains don't acknowledge the "no" they just read the word "poverty". For that reason, on my SDGs Facebook page, I tend to refer to SDG 1 as something like "abundance", "sufficiency" or "enough for all".

Professor Arvind Singhal is co-author of *Inspiring Change and Saving Lives: The Positive Deviance Way*. You can get an inspiring summary of his thinking by watching his TEDx talk on YouTube. Just search for Arvind Singhal TEDx. The essence of positive deviance is that for every challenge there will probably be someone somewhere who is providing the answers, who is making progress. So by examining, celebrating and learning from successes, we can get more of the same. That is why I think it is so important to become familiar with progress to delivering the SDGs in order to inspire more of the same.

Activity 1.4: Reframing the goals

Have a go at retitling some of the goals in any words that inspire you. Make your notes below.

The world of work

I am absolutely convinced that young people are a significant part of the answer. But there is another audience that can make a major difference. Most of us engage with the world significantly through the work we do and so getting employers on board is essential.

The truth is that the SDGs truly represent a vision for humanity that most people can and will buy into. This vision is the playing out of a fundamental instinct of all life forms: to thrive and enjoy the process. This purpose and vision will progressively and consciously become the manifesto for humanity and one that organisations will benefit from aligning to.

Just as teams within an organisation require that their team purpose and vision are aligned to those of the organisation in order to stay relevant, the same can be said at a level higher, in terms of organisations fitting into the global agenda. Failure to do so will make an organisation irrelevant and destined to decline.

Think of it this way. The world is quickly becoming aware of the urgent need to combat climate change as expressed clearly in SDG 13. Consequently nations are pledging to cut back significantly on fossil fuel usage and make more use of renewable energy sources alongside essential measures for energy efficiency. This will be a challenge for the fossil fuel industries. It is a challenge that organisations in the sector will do well to acknowledge and respond to. Those who make the transition in an intelligent manner will survive and prosper in the new clean energy age. Those who don't, won't.

So, wise organisations will choose to explore and understand the SDGs amongst other global trends. If they can consciously and visibly align to this "big agenda", they will increase their relevance in the world. This is clearly good for business, as well as being a "right thing to do".

The benefits of aligning organisations to the SDGs are good for business in many ways. Some are obvious and some are more subtle.

Obvious benefits of aligning an organisation to the goals

Some of the obvious benefits are as follows: conserving energy saves money; providing meaningful employment retains staff; promoting health reduces sickness absence; and providing life-long learning equips the workforce with skills and knowledge that benefit the operation. There are many more.

On top of this, the core business of most organisations will align to one or more of the goals. Obvious examples are pharmaceutical companies who align to SDG 3: health. Check out this photo we took on our visit to New York in December 2015 (Figure 1.2). It clearly shows the commitment of Pfizer to the SDGs – highly visible in the street-front windows of their global headquarters.

Activity 1.5: How my workplace contributes

Find more direct benefits for an organisation you are familiar with. Take a look at each goal in turn and see how many benefits you can identify from conscious and visible alignment to the SDGs. Think also about which goals are naturally supported by the services or products the organisation offers. Make your notes below:

Figure 1.2 Pfizer global headquarters New York City featuring SDGs
Source: photo taken on author's visit

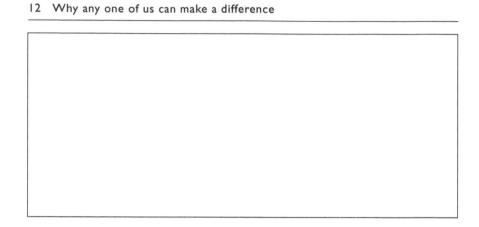

You may wish to share your thoughts with others at your workplace. Maybe run a group exercise to see what others think.

Indirect benefits for organisations

If there are many direct benefits from SDG alignment, then there are infinitely more indirect benefits. For example, asking how the products and services an organisation provides might evolve in line with the goals is a profound and relevant stimulus for innovation. Getting teams to work on projects aligned to the goals is a great way to consciously improve team working. And one of the most powerful opportunities I have been involved with is to use the SDGs as a focus for leadership development.

Developing leaders through "immersion" in globally significant projects

In over forty years at work, and half of that time specifically dedicated to leadership development, I have been fortunate to learn a great deal about leadership and how it can be significantly improved. One concept I particularly like is the difference between so-called horizontal and vertical development. Horizontal development is where leaders learn new skills for their leadership "tool kit" such as influencing skills or how to handle difficult conversations with others. Vertical development, on the other hand, is about leaders developing their mindset.

Many books have been written on the subject, such as *Immunity to Change* by Robert Kegan (Harvard Business School Press 2009). People, including those we regard as leaders, progress through various levels of psychological maturity as they grow and develop. For example, at Kegan's level three, people establish a "socialised mind" where they learn how to fit in with others in their social context. At level four there is a progression to a "self-authoring mind" where someone knows what they want and skilfully how to achieve it. The highest level, level five, is the "self-transforming mind". Here people realise that the most productive way to make things happen is to collaborate and work with others, some of whom may be culturally and otherwise very different to themselves.

Let's quickly make this relevant to the SDGs. I believe that people are most likely to move towards a self-transforming mind when they are involved and inspired by a purpose that is clearly "bigger than they are". It's easy to see, therefore, that when people are consciously working on the delivery of a project that contributes to the SDGs, they have the opportunity to embrace many perspectives and evolve their thinking for the benefit of progress. I have had the good fortune to be involved in the design of programmes that do just this. For example, my colleagues and I worked with a well-known pharmaceutical company to design a programme that took their leaders to Africa to help tackle cervical cancer. This is a clear example of using SDG 3 as a focus for leadership development.

The United Nations Global Compact

Organisations anywhere in the world, who are truly committed to the SDGs, should seriously consider signing on to the UN Global Compact. This branch of the UN believes that companies that do business responsibly and find opportunities to innovate around sustainability will be the business leaders of tomorrow. The Global Compact is a call to companies to align strategies and operations to universal principles on human rights, labour, environment and anti-corruption, and take actions that advance societal goals. Quite simply, the Global Compact is the world's largest corporate sustainability initiative and totally committed to the SDGs. In 2015, I was delighted that my colleagues at Primeast took the bold step of publicly aligning our business to the Global Compact and, in particular, to the SDGs. I will say more about the Global Compact in Chapter 9 when we examine the structures we can put in place to support SDG delivery.

SDG delivery in communities

Away from the world of work, there is so much that can be done to support SDG delivery in our communities. Again, the only limit is our energy and creativity. Community buildings and land can be used to do practical things and to educate local people. Schools and other centres can embrace energy efficiency and renewable energy. Land can be used to grow fresh local organic vegetables. This was the case in Todmorden in Yorkshire where public land, including that of the local police station, was used to grow vegetables that any local people could pick and use.

I live in the same county of Yorkshire, in a town called Harrogate in the UK. Discovering that my nearest branch of the United Nations Association (UNA) was some seventy miles or so away in Kingston upon Hull, I decided to set up a local branch in Harrogate. We now run regular meetings for people in our area to discuss challenges they wish to tackle for the SDGs. From time to time, you will see some of our meetings on the events page of the UNA (UK) website.

Activity 1.6: communities for SDGs

Think about the community you live in and the institutions you are part of. What could you do locally to raise awareness and support delivery of the SDGs? Remember, no action is too small and bear in mind the ripple effect discussed above.

SDG delivery in families

Even starting SDG conversations in our families will have a positive effect. They are likely to affect the way we run our homes, look after our health, make our buying choices and engage with local and national government – even if it's just at the ballot box. In the run up to the general election in 2015, I wrote to the local candidates to ask for a video interview. Two responded positively and expressed good support for the SDGs. Needless to say, one of them got my vote! The videos are still accessible on YouTube if you're interested.

Activity 1.7: Families for SDGs

Share your thoughts about the SDGs and why they interest you and your family. See if they will join you to make one or two initial commitments to the goals. Make a note of your family commitments below:

In conclusion

I hope this chapter has served to show how all of us can raise awareness and support the delivery of the SDGs. Whilst it is our leaders who have signed on to the goals through the United Nations, I hope you are encouraged to play a personal part rather than sit back and leave it to politicians. I hope you have identified which goals particularly inspire you and that you have already determined to take some initial steps.

Looking ahead

In the remainder of this book, I will provide you with a framework that has already been used by many organisations to support their strategic objectives. However, the framework is generic and can be used to work out how to be successful in the delivery of any purpose. In the chapters which follow, I intend to show specifically how it can be used to support the SDGs.

Chapter reflection

At the end of each chapter of *Designing the Purposeful Organization* I provided a set of ten statements to help readers reflect on the concepts I shared.
They seemed to value the opportunity so I'll do the same throughout this
book, at the end of each chapter.

Score the following statements out of ten where:

0 = not at all
2 = a little
4 = moderately
6 = mainly
8 = significantly
10 = completely

1 I have a very clear sense of personal purpose. 2 I understand how this fits within the context of the SDGs. 3 I know which of the SDGs particularly resonate for me. 4 I have a good idea about how I could personally be part of delivering these goals. 5 I have already created an initial action plan. 6 The organisation I work for or associate with has a clear sense of purpose. 7 This purpose is clearly understood in the context of the emerging global future as represented by the SDGs. 8 I have encouraged key people in this organisation to explore the SDGs to see how they could be helpful in guiding future direction. 9 I share my insights on the SDGs with other people in my family and community. 10 I have a plan to learn more about the SDGs, especially those goals that are particularly important to me.	

Chapter 2

Reasons for hope

Building on the thoughts of Al Gore

This chapter is a glimpse to the future as prompted by the thinking of Al Gore. In his TED Talk, Gore explains how technology and innovation bring solutions to climate change and sustainability in general. So, for example, advancements in solar energy are making it increasingly cheaper than fossil fuel alternatives. This will soon lead to a transformative "tipping point" and experiences with other technologies (e.g. mobile phones, tablets etc.) suggest that the rate of uptake will exceed predictions. Combine this with other advancements – such as battery performance, electric cars and driverless vehicles – and it is easy to see how quickly change will manifest. Similar predictions can be made in the fields of health and education – and even with regard to poverty, although the latter requires a significant shift in mindset and away from the libertarianism of countries like the US and (to a lesser extent) the UK. My sincere wish is that the reader will be inspired by examples great and small to play their part at home, in the community or at work.

Al Gore's optimism

With such a massive challenge as delivering the SDGs, it is inspiring to hear from respected commentators who give us optimism and hope. In this respect, the man who captured my attention in February 2016 was Al Gore, famous as a US Vice President and presidential candidate as well as for his 2006 presentation and film *An Inconvenient Truth* which warned of the dangers posed by climate change.

Activity 2.1: Hope from Al Gore

Watch Al Gore's "The Case for Optimism on Climate Change" on YouTube. What are the three questions he poses to his TED audience? Note also his reasons for optimism in the space below. Then consider some of the other SDGs and make your own notes as to why there may be reasons for hope.

Al Gore's three questions regarding climate change were:

1 Do we really have to change?
2 Can we change?
3 Will we change?

And the answers . . .

Do we really have to change?

As expected, Al Gore's unequivocal answer to the first question is a resounding "yes". He describes the fragility of our planet and how our thin atmosphere is a dumping ground for our industrial waste. He describes many causes and then focuses on the fact that we still rely on fossil fuels (a major greenhouse gas contributor) for the vast majority of our energy needs. He reminds us of the severe consequences of climate change, including storms, droughts, fires, sea-level rises and the consequential human impact including the obvious damage to agriculture and also the not so obvious migration of populations from farmlands that have become sterile. In Syria, this population migration was a significant contributory cause of the troubles currently experienced in that part of the world.

Can we change?

Al Gore describes the answer to the second question as "the exciting news". Again the answer is positive. He affirms that, time and time again, we underestimate our ability to change. For example, in 2000, it was predicted that wind generation could produce 30GW of electricity by 2010. This was exceeded and by 2015 we were producing a staggering fourteen and a half times that amount of energy. Renewable energy uptake is growing exponentially.

Countries like Germany are leading the way, recording that, on one day (26 December 2015), 81% of its energy use was from renewable sources. Solar energy is becoming a significant contributor, with costs falling dramatically and its uptake beating predictions made in 2002 for 2010 seventeen-fold. We are now set for a sixty-eight-fold increase on the 2010 prediction for 2016. This, combined with parallel improvements and cost-reduction in battery technology is good news for clean energy.

Renewable energy is described as the biggest business opportunity in the history of the world. Al Gore enquires whether there is any precedent for such rapid adoption of new technology. Well it seems there are several. In 1980, consultancy giant McKinsey forecast that the uptake of mobile phones by the year 2000 would be 900,000 users. Sure enough, in 2000, this number was sold – *in the first three days!* The total sales for that year numbered 109 million. At the time of writing in 2016, there are now 7.6 billion cell connections, more than there are people on the planet.

Why do such changes happen? First, a reduction in cost. Second an improvement in technology, functionality and quality. And finally, importantly, the fact that the developing world finds ways to leapfrog the earlier technologies and go straight to the new, saving a fortune on infrastructure costs. This is exactly the same with renewable energy. For example, with solar energy and battery storage, houses can be energy self-sufficient without a connecting network.

Will we change?

In the 2015 Paris agreement, virtually every nation in the world agreed to work together to eliminate all greenhouse gas emissions. Progress has already been made, some of which is described above. Al Gore cites further examples. China has launched a carbon market and agreed to cap emissions from six industrial sectors. The US cancelled significant mining investments and retired many existing plants, and almost three-quarters of new energy investment has been in renewables.

People power is also creating momentum and a will to change. Almost 400,000 people demonstrated in New York before the UN session on climate change and people all over the world are making their voices heard. Al Gore reminds us also of President Kennedy's promise to land a man on the moon and bring him back safely. He also affirms that the average age of the staff in the control room on the day of the moon landing was 26. That means they were just 18 when they heard the news. Promises are powerful motivators.

They've heard the promise

Most of the thousands of people I have shared news about the SDGs with are under 25. I know from first-hand experience that they have made personal commitments to change the world, to play their part in support of the SDGs. But we all need hope and encouragement and Al Gore, as ever, sets a wonderful example.

On the SDGs Facebook page (SDG2030), along with others, I post articles, videos and other media, drawing attention to the goals. Whilst it is important to describe the challenges we face, many of which are disturbing, our small team of editors tries to post a significant number of positive stories which celebrate progress towards the goals and there are many.

In the remainder of this chapter, I shall attempt to share one or two stories for each of the goals that will give us hope that the SDGs can and will be achieved. They may not be the most powerful stories; in fact I have tried to provide a balance between massive contributions and more personal contributions. This reminds us that we can all make a difference.

SDG 1: No poverty – *permaculture in Malawi*

This case study could easily have been listed under SDG 2 and indeed under several other goals such as innovation and life on land. Malawi is sadly and repeatedly ranked as the poorest country in the world. Frequent droughts, environmental degradation, poverty and arguably a colonial-bred attitude of dependency are amongst the nation's challenges. Yet, in a world that too often sees large-scale agriculture and even GM crops as the primary solution to food shortages, agents for change at the Kusamala Institute see things in a refreshingly different way. They believe that small-scale permaculture is at least part of a more pragmatic and sustainable solution. I had the good fortune to work with colleagues to facilitate a Future Search workshop for Kusamala at a Wildlife Centre in Lilongwe a couple of years ago. Their Future Search challenge was "how to create an environmentally aware and food-secure Malawi". Working with farmers, village chiefs, elders and other stakeholders, they created a plan to bring small-scale, ecologically sound farming methods to villages. One of Kusamala's instructors at that time was Luwayo Biswick. He has become a dear friend and I have watched him study and work tirelessly in his quest to end hunger. I frequently post his accounts of progress on the SDGs Facebook page.

Hear what Biswick has to say about his inspiration:

> I am inspired by the fact that in Malawi, contra to popular opinion, we have twelve solid months to grow food every year. There is never a time where we can say things are dying because it's too cold or hot. But our design has to be in tune with nature. I come from a very poor background, and have experienced long periods of hunger. My education was impeded because of poverty. These challenges strengthened my resolve to establish my own farm, a paradise where I am able to produce enough food for my wife and child, generating income for our own survival and for others who can't pay for their own food but can help

on the land. This is already happening and my land is becoming a living classroom, a demonstration to communities around us as a paradigm shift. My hope is that this work will be commended by the government of our country and able to accommodate local and international trainees. I want to see my daughter grow within the boundaries of ethical thinking governed by permaculture, earth care, people care and a sense of sharing surplus.

Building on Biswick's experience, from quite a different source, consider what happens when permaculture meets modern design and innovation.

"Huge indoor farms combining crops and livestock could transform the way we think about food, bringing exotic produce to countries such as Estonia without costing the earth." So reports *Horizon*, the EU research and innovation magazine.

> With an expected two billion more mouths to feed by 2050, the world needs to learn how to grow more food on less land.
>
> Over the last century, agriculture has intensified, dictating global diets, reducing biodiversity and contributing to global warming. With more mouths to feed every day, two recent EU feasibility studies have shown that high-yield indoor farms could be the missing link in achieving a sustainable food chain.
>
> ("Under the Dome – Multi-Storey Farms Bring Crops, Livestock Indoors" *Horizon* – the EU research and innovation magazine – 15 December 2015)

SDG 2: Food – *living salad*

This case study focuses on food waste reduction as reported in the Triodos Bank publication *The Colour of Money*. Across the UK, bags of unopened salad are tossed straight into the bin. In fact in 2013 Tesco claimed that 68% of the bagged salads they retail are wasted.

Using innovation as a potential solution, three students from Bristol University invented a novel way to reduce food waste and introduce the everyday shopper to a new kind of urban horticulture.

The "LettUs Grow" solution posed by the students is to build products that grow convenient, fully automated food for the smartphone generation, providing watering reminders and growing fresh, healthy produce within arm's reach of the plate. (See Figure 2.1.)

SDG 3: Health and wellbeing

We already have much to celebrate. In most parts of the world people are living longer and many of the killer diseases such as HIV AIDS have been reduced significantly.

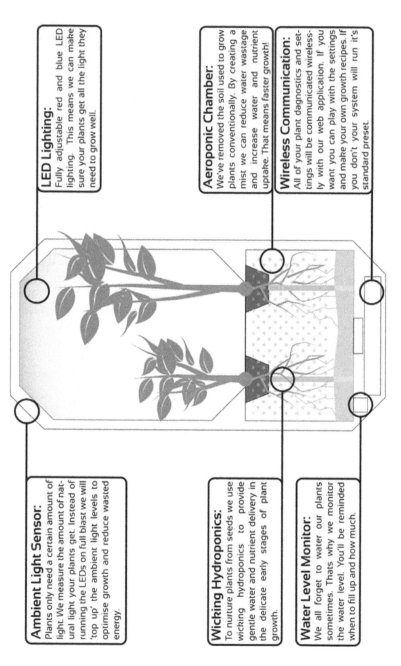

Ambient Light Sensor:
Plants only need a certain amount of light. We measure the amount of natural light your plants get. Instead of running the LEDs on full blast we will 'top up' the ambient light levels to optimise growth and reduce wasted energy.

Wicking Hydroponics:
To nurture plants from seeds we use wicking hydroponics to provide gentle water and nutrient delivery in the delicate early stages of plant growth.

Water Level Monitor:
We all forget to water our plants sometimes. Thats why we monitor the water level. You'll be reminded when to fill up and how much.

LED Lighting:
Fully adjustable red and blue LED lighting. This means we can make sure your plants get all the light they need to grow well.

Aeroponic Chamber:
We've removed the soil used to grow plants conventionally. By creating a mist we can reduce water wastage and increase water and nutrient uptake. That means faster growth!

Wireless Communication:
All of your plant diagnostics and settings will be communicated wirelessly with our web application. If you want you can play with the settings and make your own growth recipes. If you don't your system will run it's standard preset.

Figure 2.1 LettUs Grow

Source: reported in Triodos Bank publication *The Colour of Money*. More at: https://colour-of-money.co.uk/lettus-grow/

Having been a frequent visitor to Africa in recent years, I was keen to know how we can tackle malaria, still one of the biggest killers in the tropical developing world. Well it seems we are making massive progress.

According to James Gallagher, Health Editor at the BBC News Website (September 2015), in 2000 there were 262 million cases of malaria infection and 839,000 people died.

The latest report (2015) by the World Health Organization and UNICEF said malaria death rates had fallen by 60% and the cases had fallen by 37%.

They estimate that 6.2 million lives have been saved, with the vast majority being children.

In Africa, it is estimated that 700 million cases of malaria have been prevented since 2000 and it is no longer the biggest cause of death on the continent (see Figure 2.2.).

Efforts to control malaria focus on preventing people being bitten by mosquitoes and treating them once they have malaria. In Africa:

- 68% of the fall in cases was due to the distribution of a billion insecticide-treated bed nets;
- 22% was attributed to the drug treatment artemisinin; and
- 10% to spraying homes with insecticide.

Two-thirds of at-risk children around the world are now sleeping under insecticide treated nets.

So can malaria be eradicated completely? The WHO says thirteen countries that had malaria in 2000 no longer have any cases of the disease and a further six reported fewer than ten cases. This shows the disease can be eliminated from countries and potentially eradicated completely.

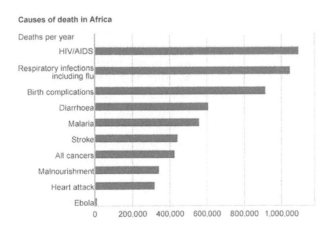

Figure 2.2 Causes of death in Africa

Source: WHO

However, sub-Saharan Africa remains the epicentre of the disease and there are still hundreds of thousands of cases on the continent each year. It will therefore require further significant progress here before talk of eradication is taken seriously.

SDG 4: Education – free online learning

I mentioned AIESEC in my introduction and my involvement with the Youth Action Summit at the UN. I recently spoke at one of their conferences in the UK about taking the SDGs into schools and universities. At the coffee break, I asked one of the AIESEC members which of the goals inspired him the most. "SDG 4: education" was his response. I asked him whether he had decided to make a commitment to this goal.

His response was inspiring. "I've decided to establish an online resource that will provide education to anyone, anywhere in the world, on any topic they wish, completely free of charge."

Almost in disbelief, I queried how this was possible. His model was simple. Experts would post online learning material and, in return, the site would sell their books and take a commission. Such a simple, yet credible, model which I hope he transforms into reality.

Of course, free online education is not new. FutureLearn, using many of the methods previously deployed by (its owner) the Open University, provides quality university courses free of charge with the only charges being for optional certification. In parallel with writing this book, I completed a FutureLearn course on climate change, completely free of charge.

Clearly this is just the beginning. Subject of course to electricity, Internet access and personal motivation, there is no reason why anyone anywhere can't have lifelong learning.

Combine these thoughts with some of the advancements in wireless technology and solar electricity. A German research institute has developed small-scale local wireless networks to provide Internet access to remote villages in the developing world. Imagine rural villages where homes and schools are equipped with solar panels, LED lighting and Internet access. The world is in transformation and, enabled by the Internet, learning is in abundance. We tend to take this for granted, especially in the developed world. However, such abundance and take up of learning has never been the case before this century and will undoubtedly accelerate progress towards the world we want.

SDG 5: Gender equality

If ever hope on this topic was epitomised in the life of one person, it must surely be that of Malala Yousafzai. Now popularly known around the world as simply Malala, this inspiring woman began her crusade at the age of 11 when she started blogging for the BBC (under a pseudonym) about life under

Taliban occupation in Pakistan. On 9 October 2012, Malala boarded her school bus and was deliberately shot three times in the head. She was seriously and permanently injured but survived and recovered in hospital in the UK.

In 2013, United Nations Special Envoy for Global Education, Gordon Brown, launched a UN petition in Malala's name, demanding that all children worldwide be in school by the end of 2015; it helped lead to the ratification of Pakistan's first Right to Education Bill.

In 2014, Malala was announced as the co-recipient of the 2014 Nobel Peace Prize for her struggle against the suppression of children and young people and for the right of all children to education. At the age of 17, she became the youngest-ever Nobel Prize laureate. She was the subject of Oscar-shortlisted 2015 documentary *He Named Me Malala*.

I will revisit the Malala story in Chapter 12, on talents and strengths, toward the end of the book.

On a very different note, but still providing hope for SDG 5, one of our UNA meetings focused on a challenge brought by a young man who was starting a career in "men's work". Convinced that men and macho stereotypes are very much part of the gender equality challenge, he seeks to provide the space for men to explore more transformational ways of being that are authentic, wholesome and different from the norm which would otherwise perpetuate cycles of inequality and potential violence.

SDG 6: Clean water

There have been many successes in the efforts to bring safe, clean water to almost a billion people who still don't have such access, especially in the developing world. Here in the UK, people are very familiar with the work of Water Aid which claims to have reached 23 million people in 37 countries since 1981 in the quest for safe water supply and sanitation.

However, in terms of hope, I'd like to focus on some of the fantastic technologies that are emerging and combining to provide innovation for water supply as well as meeting other needs along the way. As with all innovation, maybe some of these ideas will be short lived, but inevitably, some will become the game-changers the world needs to deliver on goal six.

So, just as an example, let me introduce you to Watly, a new company using modern technologies to deliver breakthrough solutions. Based in Italy and Spain, a number of bright young innovators have developed a device that purifies up to 3 million litres of water per year powered by solar energy.

Watly generates the very same energy it needs to function, so it does not require fuel or a connection to the electricity grid. It efficiently desalinates ocean water, eliminates all pathogens and microorganisms from previously polluted water, including: viruses, bacteria, parasites, fungi or cysts. It removes inorganic compounds as well as poisons: arsenic, benzene, heavy metals (such as lead), chlorine, chloramines and radionuclides. It purifies water from any organic compounds and liquid contents of latrines. It does not need membranes or filter substitutions. It even purifies radioactive water. The physical principle underlying this innovative system is called vapour compression distillation. It is by far the most effective and powerful method of water purification and desalination available. To find out more, watch their tutorial videos at the Watly website.

SDG 7: Energy

We've already said quite a bit about goal seven in the preamble to this chapter, referencing Al Gore's thoughts about the rate of progress. So here, I want to emphasise how easy it is for any individual to impact this goal. There is almost no limit to the actions that any of us can take to save energy and to make sure the energy we use is clean and renewable. In our homes, we can first of all choose to purchase our electricity from a renewable energy supply company. Based in UK, when I first became interested in the SDGs, a friend introduced me to Good Energy, an energy supplier that provides electricity only from renewables. I became very interested in their work and as well as signing on to their service, I also chose to invest a small amount in the company.

In January 2017, I had the good fortune to interview the CEO of Good Energy, Juliet Davenport, for a podcast (available on my LinkedIn page). Juliet is an inspirational and purposeful leader who established Good Energy as a result of her concern for climate change (SDG 13). The company not only retails electricity from 100% renewable sources but also sells clean gas from non-fossil sources. They also provide support to users seeking to install local generation systems using a range of renewable technologies. Juliet described the benefits of aligning the work of her business to the SDGs. It ensures global relevance and provides inspiration to all stakeholders. To find out more, just listen to the podcast or watch the plethora of inspiring videos featuring Juliet on YouTube and at the Good Energy website.

As well as interviewing Juliet at their offices in Chippenham UK, I also got the chance to run a workshop on the SDGs with their staff. I thoroughly enjoyed the experience and you can imagine my delight when, just a few months later, Good Energy rebranded and placed the SDGs at the heart of their offering. At about the same time I installed solar panels and batteries at my home in Harrogate and chose Good Energy to administer the associated Feed-in Tariff which pays small scale generators for the electricity they produce.

By the way, how we invest our money, be it for our pensions or any other purpose can support or detract from the SDGs. I also chose to invest in renewables through Triodos, a Dutch ethical bank with offices in the UK. One of the advantages of renewable energy is we can all become investors in this great emerging industry as well as generators through solar panels or even wind turbines.

Then of course there is consumption. In our homes, we can invest in energy efficiency measures such as insulation and LED lighting. Every little helps.

The convergence of technologies on the energy front is really exciting, especially in emerging economies. Having spent a significant time in Malawi in recent years, I know the difference a solar panel, combined with LED lighting, USB charging and Internet access can make. Add this to the permaculture solutions discussed under goal one above and village life in the beautiful African countryside can be transformed without the need for a national electricity grid.

I could arguably talk about transport systems here or indeed under goal thirteen, climate change. Clearly the move to electric transport systems is a major contributor to a world of clean energy, but let's save that for our discussions on innovation and clean cities.

SDG 8: Decent work and economic growth

There are so many reasons to be hopeful for the delivery of this goal. Many of the goals already discussed can bring new and improved employment prospects. For example, permaculture is both interesting and rewarding work. This small-scale farming approach close to home is healthy in terms of the manual work and in the production of healthy organic food. It removes the need for long distance travel to work and working with, rather than against, ecosystems takes a lot of the back-breaking labour out of play.

The technologies discussed above also provide employment opportunities at all stages of the pipeline, from research through manufacture, distribution and implementation, and even in maintenance and recycling. Take the installation of solar energy systems. The US Department of Energy recently published data on the growth and trends in "solar employment". Between 2010 and 2016 the numbers employed in installation work had grown by over 200%.

But the reason for hope I'd like to focus on here is a bit less glamorous and yet it is something we can all impact. It is our personal purchasing power. According to the Ethical Consumer Markets Report 2016, 53% of the UK population is choosing to avoid buying products and services over concerns about ethical reputation. In addition, local shopping for ethical reasons grew significantly in 2015 with consumers increasing ethical spending in their communities by 11.7%.

People are clearly becoming more demanding when they purchase their food, clothing and other goods. There is much we can do to check out the ethics of our suppliers. We can research their ethics at their websites, ask probing questions at the check-out and consequently modify our buying habits.

One very powerful associated question is whether an employer is a member of the UN Global Compact (more of this later) and thus signed to principles of human rights including fair employment practices. Even asking this question of our suppliers might cause them to find out more.

SDG 9: Innovation and infrastructure

We've discussed above several innovations with food, energy and other systems. But when I think of hope regarding this goal, I think of young people. Indeed, many of the innovations discussed above are the brainchildren of young people. So many young people demand employment that is more than just turning up to work for the pay-cheque.

When I was young, the developing world was something I had read about in newspapers or seen on television. My first trips to Africa, Asia and South America were all to do with the work I was doing in my 50s. But today, many of the young people in the so-called developed world have travelled extensively. They have met and stayed with people who struggle with poverty and the other challenges represented by the SDGs. Many of them are not willing to ignore the injustices in our world.

It is for these reasons that I have immense hope in our young people. Fuelled by a hunger to make a difference and high standards of education, young people in the developed world are prepared to collaborate with their brothers and sisters in less fortunate situations, to explore and resolve problems using innovation appropriate for the context.

This is why some of the most satisfying work I do is going into schools and universities to meet and talk with young people. They are always inspired by the SDGs and many choose to align their studies and career choices to the better world the goals present. This is a massive job for all of us and there will be some of you reading this book who are keen and eager to play a part. If you would like to be involved in taking the SDGs into education, get in touch and I'll be happy to provide tips and resources.

Activity 2.2: The Venus Project

For those who are inspired and take hope in innovation that borders on sci-fi, I encourage a visit to YouTube to watch some of the videos relating to The Venus Project, the brainchild of Jacque Fresco. In a world where technology has minimised manual work, where automation and artificial intelligence prevail, The Venus Project proposes that people will be heavily involved in learning, research and the arts. People don't own anything, and there is no money. The project isn't just a flight of fancy, it has been prototyped in self-sufficient communities already. Whilst, I don't envisage a full-scale transformation to a Venus Project world anytime soon, I do believe that many of the principles will begin to play out and contribute to the delivery of the SDGs. What do you think? Watch the videos and make your notes below.

SDG 10: Equality

This is another of those goals that talks about eliminating a negative. It focuses on reducing inequality within and between nations. So, taking a positive perspective, this is about equality. Gender equality is covered in goal five, so this goal is about the gap between rich and poor within nations and between wealthy and poor nations.

In many countries such as the US and here in the UK, we seem to be heading in the wrong direction with the gap between rich and poor widening. US Census data (available easily on the Internet) shows clearly how this gap is increasing. So where indeed is the hope?

This is probably one of the hardest goals to make progress on and it is principally where people power must come into play. We discussed above how purchasing power counts when it comes to employment practices. There is a strong link here to equality. Do we choose to buy from companies that exploit their workers at home or abroad? Do we support companies who pay their leaders disproportionately high salaries and bonuses? Do we think twice before doing business with those who avoid paying taxes that might otherwise contribute to social improvement?

What about our involvement with government? Do we lobby for justice? Do we vote for selfish reasons or for equality in the world?

My reason for hope in this goal is a combination of young people, education and social media. This won't be a quick fix. As I visit schools to talk about the SDGs, some teachers are enquiring if they could do more to align their curriculum to the SDGs. My response is that the goals are a vision for humanity so, "Yes." Any school, university, business or political party can choose to align its curriculum, strategy or manifesto to the SDGs. By doing so, those involved will become more aware of inequality and more likely to do something about it.

I also wonder whether the SDGs present a unique opportunity to influence the politics of our world. I personally wonder whether the SDGs could be adopted in full as the manifesto for a political party. It seems to me that all the questions that get posed at election time are indeed answered in the SDGs, especially when it comes to equality.

Activity 2.3: Global alignment

Do the SDGs offer opportunity to begin a global alignment of political thinking? I don't know of this happening anywhere in the world and I'm genuinely curious about what your thoughts might be on this topic. Make some notes below and please let me know what you think.

SDG 11: Sustainable cities

I started my working life as an engineer, so maybe that's why I find this goal so exciting and a reason for lots of hope. We touched on clean energy innovation above so I shan't revisit that, except to say that when cities go all-electric, sourced from renewable generation, the benefits are significant. They become clean, efficient and simply nice places to live.

At one of our UNA meetings in Harrogate we invited Tom Riordan, the CEO of nearby Leeds City Council, to share his plans for a clean city. A key area was transportation, with the introduction of electric vehicles to the council's fleet of service vehicles. Leeds has already seen the pedestrianisation of much of its city centre and there is more to come. Eventually petrol and diesel vehicles will not have access to the city centre at all.

Thanks to the pioneering work of entrepreneurs like Tesla's Elon Musk, electric cars are advancing rapidly and when this technology is combined with that of driverless vehicles, the impact on our cities and suburbs will be significant. It is likely to prompt fundamental behaviour change. How many years before people cease to own their own cars and instead call a driverless pod on their smart phone and release it when they've made their journey? Some futurists believe this technology could result in 80% fewer vehicles on our roads.

Some of the most powerful changes are the most simple and are already happening. In 2016 I traded my fuel-thirsty Harley-Davidson motorcycle in for an electric push bike. I now find I can get around my home town quicker than I can in the car, it isn't hard work and I get some gentle exercise in the process. With many cities improving cycle lane provision, something as simple as choosing e-bikes (or normal cycles of course) can be a game changer.

SDG 12: Responsible consumption

There is significant overlap between the various SDGs, which I personally find inspiring. So, unsurprisingly, the topic of responsible consumption has been woven into some of my reasons for hope with the other goals. Choosing organic local food and renewable energy, purchasing products from ethical companies or riding a bike instead of driving are all facets of responsible consumption.

The innovation that specifically gives me cause for great hope regarding this goal is the advent of the sharing economy. Supported by Internet-based applications, we can now make sure our property is put to good use through sites like Airbnb. We can speak with friends on Skype without consuming fuel and recycle our waste through local services.

But the reason for hope that has inspired me the most is Freecycle and other Internet applications that allow us to post adverts to recycle items we no longer have use for so someone else can make use of them. And of course we can also look for items we need without always buying new.

I recently had a clear-out of my garden shed and offered two children's bikes, a couple of bike racks and an old Victorian cast iron fireplace that had been cluttering shed space for far too long. The items were gone within a day and their new owners were delighted. I also felt good about being able to help them. It's even possible to check out recipients before agreeing to let them have your goods, just to make sure they are indeed going to a good home.

The reason this is such a reason for hope is that it represents a change in mindset. Are we gradually entering a world where we stop thinking of our possessions as "mine" and start thinking of them as a shared resource? Is this the sort of world John Lennon envisaged when he wrote the powerful lyrics to his song "Imagine"?

SDG 13: Climate

For me, this is the big one. It is inextricably linked to most of the other goals, but it is also the one where the clock is most profoundly ticking. There was a time when the number of people apathetic to climate change vastly outnumbered those who cared. But the balance has shifted significantly. Now, most people know this is something we need to attend to, even if they haven't done their homework to see just how urgent it is.

If we don't act decisively, we will see a rise in sea levels that will threaten most of our major cities and we will become victims of catastrophic collapses of ecosystems with the associated demise of planetary diversity and our own food systems.

So where is the hope? This is actually where we came in to this chapter. Al Gore has been leading the fight to mitigate climate change and it's encouraging to note that Barack Obama has been focusing his attention on this topic since his presidency. Much of my hope is founded in the inspiring thoughts of Al Gore that we discussed at the start of this chapter. I have little to add except to say that every little counts.

Inspired by Gore, I have changed my energy supplier, committed to no more fossil-fuelled vehicles, and installed solar panels and batteries at my home. I've shifted my pension investments to support the renewable energy industry. I didn't make all these changes overnight, but, strangely, they happened very soon after watching Al Gore's inspiring videos. So why not watch just one of his videos as a starter and see where it leads.

SDG 14: Life below water

When we stop to reflect on the vastness of the seas and oceans that cover the surface of our planet it's hard to feel anything but awe and wonder. Despite this, mankind has a devastating impact on this vast habitat. We kill endangered species such as whales and coral; we overfish and upset the ecosystem in ways that impair sustainable fishing; and we dump our waste, including plastics, into the water.

So where is the hope? Once more, I look to the power of young people, technology and innovation. I also look to agreement between nations to legislate against unsustainable marine practices.

I briefly mentioned earlier one of the attendees at a workshop I ran at a UK university who connected his passion for scuba diving with the need to clean the seas of plastic. Shortly after meeting this inspiring man, I read about 19-year-old Boyan Slat who, whilst still at school, launched a project to create an Ocean Cleanup Array that could remove 7,250,000 tons of plastic waste from the world's oceans and make them available for recycling. His paper won several prizes including Best Technical Design 2012 at the Delft University of Technology. Boyan continued to develop his concept and he revealed it at TEDxDelft 2012. He subsequently founded The Ocean Cleanup Foundation, a non-profit organisation which is responsible for the development of the technology.

SDG 15: Life on land

We live in a paradise, thriving with countless species of life. As a supposedly "intelligent" species that is part of this ecosystem, we are able to choose whether we have a positive or negative impact on our habitat and the vast array of life it supports. Every day our social media feeds show us the worst and best of who we are as a species. We see the brutal killing of endangered species for sport and so-called fun and most of us react in disgust. We see communities who organise themselves to plant new forests and we applaud. In all of this we get to choose. We can choose to watch from the sidelines or to do something. We can choose whether to buy meat from industry associated with forest destruction. We can choose free-range eggs over factory farming. Every day we get to choose and if we are wise we act as humble participants in this adventure we call life, seeking to leave the world in better shape than we found it.

My hope for this goal is founded in the amazing people who have sought to focus and share the beauty of our planet. Iconic figures like Sir David Attenborough. Organisations like National Geographic and ordinary people like my friend Jane Keogh who, at one of our UNA meetings, sought to combine her passion for education, her heart-felt love of endangered rhinos and her connection with Zambia. To cut a long story very short, Jane tells me that she wouldn't have made an intentional trip to become involved in an educational project in Zambia had it not been for the encouragement and creativity of her friends at UNA when she brought her challenge to our meeting in 2016.

SDG 16: Peace and justice

I am often asked which of the SDGs I care most about and I struggle to answer. I love my planet, the life it supports and my fellow humans who I get to do life with. As an engineer, the energy and innovation goals fascinate and intrigue me, especially where they mitigate climate change. But the goal that touches my heart in the way that no other can is SDG 16 – which I refer to as simply "world peace".

Quite frankly, the injustices inflicted on humans by other humans shock and trouble me deeply. When I see war-torn cities, death of innocent people and those fleeing their countries, the tears well up inside.

Again I ask, with desperation, where is the hope? And yes, I do sense some answers. However tenuous you may consider them to be, surely the way forward is to follow the hope as our hearts discern.

There is hope in the increasing dialogue between nations, such as at the United Nations and the unions of Europe and Africa.

There is hope in the work of those seeking to encourage and enable peace through mindfulness practices including prayer and meditation. The work of author Greg Braden and the HeartMath Institute springs to mind.

But of all the things that give me hope, the one that inspires me most might seem a little off the wall. It is maths, financial accounting to be precise. The stark statistic that I cannot get out of my mind is that the budget for dialogue at the United Nations is (according to one source) less than 2% of what we spend on arms globally. This really gives me hope, especially as most people would agree that investment in dialogue, education and getting to know people from other cultures better is a significant precursor to peaceful society, one that ultimately cuts our need to spend on arms. So my sense and my hope is that this is very affordable if we can fight against the politics of fear.

We don't have to look very hard to see the dangers associated with the politics of fear. In 2015, in my own country we have seen it with Brexit and decisions to spend money on Trident. In the US we see the rise of fear strategies that are almost unbelievable and significant numbers of the population buying into them, despite lessons learned from the past that led undeniably to two world wars in the twentieth century.

So why hope? My friends regard me as an optimist and I really do believe that the politics of love will prevail over the politics of fear.

The truth is they already are. Despite popular news and fear mongering by interested parties, the truth is we live in a far more peaceful era than has ever been the case. The man who gives me the greatest hope in all of this is Harvard University cognitive neuroscientist Steven Pinker in his book, *The Better Angels of Our Nature: Why Violence Has Declined.*

Just a couple of centuries ago, violence was pervasive. Slavery was widespread; wife and child beating were acceptable practices; heretics and witches were being burned at the stake; pogroms and race riots were common; and warfare was nearly constant. Public hangings, bear-baiting and even cat burning were popular forms of entertainment. By examining the evidence, anthropologists have found that people were nine times more likely to be killed in the tribal warfare of those days than in the killing and genocide of the war-torn twentieth century. The murder rate in medieval Europe was thirty times higher than today.

In over 850 pages of data and analysis, Pinker identifies the institutional and cultural changes that have led to this dramatic reduction in violence in our world. If you don't have the patience to read the detail of this evidence, as is the case with so many of the world's leading researchers, Pinker is prolific on YouTube. You can choose from several clips of just a few minutes to complete recordings of his conference seminars. No one has inspired me more to believe that world peace is possible than Steven Pinker.

SDG 17: Partnership

According to the UN Sustainable Development Knowledge Platform (a vital resource available online for all advocates for sustainability), this goal is described as follows:

> Achieving the ambitious targets of the 2030 Agenda requires a revitalized and enhanced global partnership that brings together Governments, civil society, the private sector, the United Nations system and other actors and mobilizes all available resources. Enhancing support to developing countries, in particular the least developed countries and the small island developing States, is fundamental to equitable progress for all.

So, in a nutshell, it's about us all working together in a cohesive and conscious way to deliver this powerful agenda. This is exactly why the subtitle to this book is *The Sustainable Development Goals as a Blueprint for Humanity*. The more we consciously align to the SDGs in every aspect of human society from the United Nations, through governments, education, communities and right down to each one of us, the more chance we have of delivering this powerful agenda. Company by company, school by school, household by household we can do it. I sincerely hope you will take some of the logic from my previous book, *Designing the Purposeful Organization* re-presented in the coming chapters and apply it to the quest of delivering a world suitable for future generations.

Chapter reflection

Score the following statements out of ten where:

0 = not at all
2 = a little
4 = moderately
6 = mainly
8 = significantly
10 = completely

1 I know which of the SDGs particularly inspire me.	
2 For each of these goals I can cite at least three case studies that give me hope that the goals will be delivered.	
3 I have shared these case studies with others in conversation.	
4 I have made direct contact with people involved in these case studies to offer my support and practical help.	
5 I have found information on these case studies and shared it on social media.	
6 My curiosity has led me to find out more about progress in these matters.	
7 I have asked my friends and family about the hope they have for the SDGs.	
8 I take every opportunity to counter pessimism regarding the SDGs with good reasons for hope.	
9 I monitor news feeds and share reports of progress towards the SDGs to inspire others.	
10 I have consciously used cases for hope to prompt personal action on my part.	

Chapter 3

A personal story
The author's journey to now

This chapter provides context for the reader, explaining the author's journey into the subject of sustainability and his decision to become an advocate for the SDGs and to chair a branch of UNA dedicated to this purpose.

Why I wrote this chapter

I've taken a bit of a liberty with this chapter by writing a short biography about myself. It's not that I want to blow my own trumpet or make this book a personal affair. I just think it's relevant for you to know a bit about my life at headline level in so far as it is relevant to the SDGs.

We all have our own personal stories to tell and by telling you how I came to be writing this book, I want to encourage you to write the headlines of your own story, perhaps at the end of this chapter. Clearly you have an interest in the SDGs or the general topic of sustainability. How did that come about? It may help you make sense of your life's journey and how your interest has arisen if you spend a little while thinking about it.

Anyway, it is your choice whether you wish to read this chapter. If knowing about the author of this book is less relevant for you than the topic of the SDGs and learning how to design a purposeful world, then you will lose little by skipping this short chapter. I shan't be offended and it won't interrupt your flow in any significant way. If you just want the headlines, then simply skim through the bold headings below without reading the "small print".

Dreaming of a better world

For as long as I can remember, I have dreamt of a better world, a world where everyone can thrive and be happy. This is not just about having a better world for humanity, but rather a better world for all of life. For all of life is dependent on all of life. When one species or race exploits or abuses another, things start to go wrong. When we put our personal interests above the interests of the wider system, there may be perceived short-term gains but ultimately we all lose out.

Born and bred in industrial Yorkshire

I grew up on the outskirts of Leeds, in the industrial heartlands of West York-shire in the UK. My senior school bordered on the local foundry and there was a coal-fired power station just around the corner. There were just over a thousand children at Cross Green Comprehensive School, drawn from a range of backgrounds. Inner-city kids mixed, not always peacefully, with those from the wealthier suburbs. The mainly white kids weren't as welcoming as they might have been to the black minorities. I remember feeling the tension, usually keeping my head down, but occasionally sticking up for the bullied as best as I could.

Straight from school to the world of work at 16

In 1972 I left school at the age of 16, determined to enter the world of work. Enjoying science, it was a choice between the foundry, the Royal Air Force (where my Dad had served during the war) or the local electricity board (where our neighbour worked). I chose the latter and trained, qualified and worked as an electrical engineer for thirteen years before retraining in health and safety. I took this opportunity to study for a master's degree in safety and reliability. I had a say over my choice of modules and I recall being most interested in environmental management and total quality management, with a focus on the human factors. My dissertation was on values and behaviours in the workplace.

Big changes in the UK electricity industry

In the 1990s the electricity industry faced massive change and I found myself serving as a project manager and then as a programme manager. The purpose of my particular change programme was to introduce competition into electricity metering. I had over twenty projects running simultaneously and, after a year or so, the programme was audited. My team scored well with the exception of one facet: culture change. I was informed that no matter how good we were at managing systems and processes, if we ignored the dynamics of people and their values and behaviours, the change would simply not work.

An interest in cultural change in organisations

I became fascinated and, in the late 1990s, visited a company called Human Synergistics in Detroit, US. They taught me how to measure current culture, the culture needed to support a change in direction and the leadership styles that would role model the way. I became accredited in their methods and also created a model to remind leaders to maintain the balance between structure and culture. I affirmed that both were necessary in order to deliver any outcome and that the outcome in turn must serve a clear purpose.

Connecting to the importance of purpose and vision

At that time, purpose and vision were not in everyday management speak. Purpose especially wasn't particularly well understood. It simply made sense to me that ultimately everything and every endeavour serves a purpose. Therefore purpose should be the anchor for everything in an organisation.

Becoming a consultant-facilitator at the start of the new millennium

On leaving the electricity industry at the turn of the millennium, I took my experience into the world of consultancy. In 2002, I was invited to lead a consultancy in Harrogate, UK, called Primeast specialising in leadership, organisational development and team work. The model I drew in my previous role became renamed PrimeFocus (you can find it at the start of the next chapter) and slowly it became a growing part of what my colleagues and I offered.

Writing about purposeful organisations

I wrote and published a number of handbooks on PrimeFocus and associated concepts round about 2007. In 2014, our ideas "came of age" when publisher Kogan Page asked me to write *Designing the Purposeful Organization: How to Inspire Business Performance Beyond Boundaries* (affectionately known as "DPO" and referenced as such throughout this book). The book was published in February 2015 and has enjoyed early success. I shall provide an overview of the book in the next chapter but for now, suffice to say, researching and writing the book took my understanding of purpose to a whole new level.

Connecting to the idea of a purposeful world

In December 2014, as I was penning the final chapter of DPO, coincidentally or otherwise, I discovered the draft Sustainable Development Goals (SDGs). I became inspired and engrossed. In the closing paragraphs of DPO, I started to ask "What next?" I sewed a few seeds and talked speculatively about the idea of "a purposeful world".

Committing to the SDGs

Over the Christmas period in 2014, I made a commitment to devote as much of my energy as I could to the promotion of the SDGs. To me, at that time, the idea that world leaders might sign up to seventeen goals that together formed the greatest shared commitment humanity had ever made was simply awesome. And if 2030 saw the delivery in full of this vision, I resolved

I would be able to die a happy man. However, I'm hoping I make it to my 75th birthday still in a fit state to party!

Spreading the word and joining UNA

I didn't know where to start but I just wanted to tell as many people as I could about the SDGs. I started with a few talks in schools and universities and also formed an "action learning set" which met at our company's office. In the summer of 2015, I learnt about the United Nations Association (UNA) and decided to join. The nearest branch of UNA was on the east coast of UK in Kingston upon Hull, some seventy miles away. I discovered that there used to be a Harrogate group that had disbanded so I made a bid to UNA to restart the group and devote it entirely to the promotion and celebration of the SDGs. I also set up an SDGs page on Facebook (SDG2030) to inspire and celebrate progress towards the goals.

In the US when the SDGs were formally adopted

In September 2015 I was speaking at the Annual Conference of American Talent Development (ATD) in Houston, Texas, on the topic of DPO. At the same time, in New York, world leaders did indeed formally sign up to the SDGs. I was overjoyed, if a little disappointed at the lack of mainstream media coverage of what is surely the biggest human commitment on our planet ever.

The ripple effect: one thing leads to another

One thing I have learnt about my SDG journey is that, when you commit to something, the universe conspires to make things happen. The bar gets raised. In December 2015, I was out on a walk in Harrogate with our company's founder, John Campbell, talking about sustainability. John is also an enthusiast and the author of *Ignition: Creating a Better World for Our Grandchildren*. As we strolled through Valley Gardens in Harrogate, John's phone rang. It was Phil Clothier, a dear friend of ours and the CEO of Barrett Values Centre (BVC). John stopped walking and turned and looked at me. "I'm with him now," he said to Phil. After a few minutes, John asked me if I was busy later that month, which I guessed I probably would be but asked why.

Working at the United Nations in New York

John explained that Phil was inviting the two of us to the United Nations in New York to facilitate some workshops at the 2015 Youth Action Summit being held there. Incredibly, John and I managed to shuffle our diary commitments and book some last minute flights.

At the summit, we facilitated a number of workshops for young leaders from AIESEC International, the world's largest youth-led organisation that develops young people in leadership through placements and engagement with leading businesses.

From ripples to tidal waves

AIESEC had committed to take the SDGs to "every young person everywhere" but needed to develop a method. When they discovered this was something I had been doing, they asked if, after the workshop we were working on, I would be able to offer training to 400 of their young leaders in Hall 1 at the UN. The outcome of that part of the AIESEC journey was testified earlier in this book.

Starting this book

I was pretty busy at the start of 2016 and my enthusiasm for the SDGs was pulling me into several speaking engagements and workshops. On top of this, I was still working hard to engage audiences and my colleagues in the methods set out in *Designing the Purposeful Organization*. In May I bumped into Rebecca Marsh from Greenleaf Publishing at a conference in Harrogate. We discussed my idea for *Designing the Purposeful World*. She was most encouraging and we consequently met at Greenleaf HQ in Saltaire the next month. I started writing immediately.

Activity 3.1: My journey to "now"

Having read my story to the point I started writing this book, take a moment to reflect about your own journey to "now". Perhaps jot some notes, including the big headings in the space below.

Chapter reflection

Score the following statements out of ten where:

0 = not at all
2 = a little
4 = moderately
6 = mainly
8 = significantly
10 = completely

1 I have considered my life journey in the context of the SDGs.	
2 I can point to a number of personal achievements that have already made a positive contribution to the SDGs.	
3 I can draw on personal experiences that have motivated me to get more involved in the SDGs.	
4 I can recall places where I have travelled that have reinforced the urgency of the SDGs.	
5 I can think of several people whose contributions to the SDGs have been significant and who have influenced me in a positive manner.	
6 The work I have done in the past has contributed to the delivery of the SDGs.	
7 The work I do now contributes significantly to the delivery of the SDGs.	
8 I am consciously steering my career going forward to contribute more to the SDGs.	
9 I know how my own actions cause positive ripples of action in others.	
10 I can relate examples of where this has positively influenced delivery of the SDGs.	

Chapter 4

Designing purposeful organisations

A synopsis of the necessary conditions for change

This chapter is a synopsis of *Designing the Purposeful Organization: How to Inspire Business Performance Beyond Boundaries*. The book outlines eight necessary conditions for change: purpose, vision, engagement, structure, character, results, success and talent. Purpose is described as the reason for being for any organisation – the "Why". However, it can shift significantly dependent on who's observing, so stakeholder perspectives are vital and must be aligned. From this work, the inspiring purpose can be derived and can be used as the focus for everything the organisation does. Vision is the manifestation of purpose at significant time horizons and can be considered the destination for the change. It gives people a common goal or set of goals. For a vision to manifest there must be engagement with everyone involved. Consequent to successful engagement are the structures that support the delivery of the vision. This is everything from plans, teams of people and communications structures to buildings and systems. Alongside the structures for change there has to be character – the necessary values and behaviours that get people involved and ensure they work together effectively. Results measure progress to the vision and take the form of various key performance indicators, whilst success is the felt sense of achieving something worthwhile. Results and success are therefore different and both are necessary. Finally, nothing happens without deploying the talents of those involved. A successful change will tend to play to the strengths of everyone.

Creating the conditions for change

A few years ago, I coined the phrase *Leaders see a better future and create the conditions in which it will happen.* My colleagues and I have since used it as a modern definition for leadership, quite different to previous thoughts, which typically were about *leaders making things happen through others*, words which had an old-fashioned autocratic flavour to say the least.

Building on my experiences in the electricity industry, in the late 1990s I described these "conditions for change" in a conceptual framework, later to become known as PrimeFocus (see Figure 4.1).

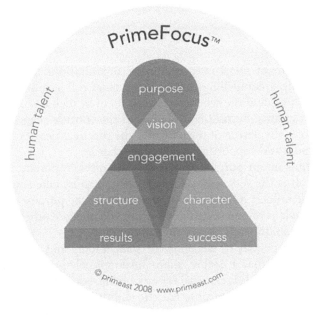

Figure 4.1 PrimeFocus™
Source: with kind permission of Primeast Ltd

Designing the purposeful organisation

In 2015, Kogan Page published my book *Designing the Purposeful Organiza-tion: How to Inspire Business Performance Beyond Boundaries* (DPO). It was mainly aimed at corporate human resources professionals. I provide a sum-mary in this sequel, *Designing the Purposeful World* (DPW), simply because the principles for designing purposeful organisations can be extrapolated and made accessible to us all as we strive to tackle the larger scale challenge of making the world a better place. The logic for doing this is explained in the next chapter which touches on the principles of self-similarity or fractals.

DPO is the culmination of nearly thirty years' service by facilitators at Primeast for their clients in the fields of leadership, organisational effective-ness and teamwork. It is principally aimed at professionals in organisational development and the leaders they support. I collectively refer to this target audience as the "organisational architects" who craft our modern-day work-places. It is in this book that the "conditions for change" are discussed, one by one in eight corresponding chapters: *purpose; vision; engagement; structure; character; results; success;* and *human talent.*

Similarly, in DPW, we explore these eight conditions but they are written as they play out in the quest of creating a purposeful world. But for now, let's concentrate on DPO so you can understand the basic concept and will have a better feel when we come to discuss the eight conditions in DPW.

Purpose

The principle aim of DPO is to encourage organisations to spend quality time establishing a compelling purpose that takes full account of all stakeholder perspectives and the synergy between them in such a way as to inspire business performance beyond boundaries, as the subtitle suggests. The first chapter explores the nature of purpose and the practicalities of establishing a primary purpose and supporting stakeholder charter.

The key thing about purpose is that it is the most important asset of any organisation. It is the very reason for existence. It must take account of and inspire all stakeholders. Organisations who only see the purpose of what they do through the lens of one stakeholder group will alienate other stakeholders resulting in instability and, ultimately, failure.

The other key benefit of engaging all stakeholders and getting to know what they value is that a deeper, more inspirational, purpose will arise to motivate all parties. This is essential for any organisation, whatever the sector.

A clear and compelling purpose has the ability to unite, align and inspire people. This alignment also improves efficiency by ensuring that all parties pull in the same direction.

Vision

Chapter 2 is about vision. It stresses that, whilst purpose provides focus for the organisation, a clear vision informs where it will be at the most important time horizons. I explore how a vision arises, how it can be shared and adapted for different audiences and how people can contribute to an evolving journey. There are several practical tools and methods for enabling leaders to work better with vision.

In DPO, I draw the distinction between a snappy "aspirational vision" such as *being the world's best widget-maker* and a broad, comprehensive vision which has the power to enable all parties to "see" the future. After purpose, which can usually be expressed succinctly, a clear vision should be the next most powerful motivator for progress.

Engagement

Leading naturally on from the purpose and vision, Chapter 3 of DPO examines the nature and importance of engagement. I draw on published evidence to support the business case for engagement. Then I share learning from the natural world, including how cells learn and evolve in the context of our bodies and how this can inform our engagement strategies. I look at the benefits of engagement that supports the development of a learning organisation, indeed a purposeful learning organisation. The concept of a learning ecosystem that mimics natural engagement is proposed as a design template.

True engagement, as opposed to one-way communication, is one of the key methods to ensure distributed leadership. When people get to work things out for themselves and become part of the solution, significant progress is made. Whilst powerful in organisations, this factor is even more important when it comes to creating a purposeful world.

Structure

In Chapter 4 I demonstrate how to align all the structures and processes of the organisation specifically to the compelling purpose. I extend the excursion into science with an introduction to fractal geometry which demonstrates that efficient structures at one level will play out in a self-similar manner at other levels as we zoom in and out of the organisation. I shall say more about fractals in the context of a purposeful world in the next chapter of DPW. I also stress that, whilst organisations in the modern world are naturally *complex*, they should not be overly *complicated*. This reinforces the need for simplicity and paying close attention to matters such as the dismantling of redundant systems.

All ventures require structure, but that doesn't mean that the "corporate centre" has to create everything. With an organisation as vast and complex as humanity, the players (all of us) can usefully examine the structures, processes and policies available to us and take, adapt and create a mechanism that is simple, useful and gets things done without being bureaucratic and stifling. This is another reason why, when it comes to changing the world, ordinary people with distributed leadership can achieve things that governments cannot.

Character

Chapter 5 looks at the other side of the coin from structure as I demonstrate how we can manage the character of our organisation. I examine the power of culture from both geographic and corporate perspectives. The relationship between culture and performance is evidenced and we examine a number of the most popular and relevant diagnostic tools for measuring and managing culture. As in other chapters, there are several mini case studies and personal stories that bring the topic of culture change alive for the reader.

Results and success

Chapters 6 and 7 explore two contrasting outcomes from the purposeful organisation which are both essential in their own right. First, under the heading of "results" I stress that the vision should define what needs to be measured and how this might be done in such a manner that tracks progress towards its delivery. I also emphasise that results, whether good, bad or indifferent should inform the decision-making process. Success on the other hand is both an outcome and a felt sense, one that is essential to motivation and ensuring performance. I also propose a method for shifting the felt sense of success from one that is personal to one that is shared by a team. In this book (DPW), I have dealt with results and success combined in one chapter.

Human talent

In Chapter 8, I move to the topic of talent management and describe an eight-step approach to developing and implementing a progressive talent management strategy. Such an approach recognises values, and develops and uses the unique talents of everyone in the delivery of the organisation's purpose. It is thereby inclusive and incorporates best practices such as playing to people's strengths. Included in the eight steps are the essentials of leadership and teamwork, non-negotiable disciplines yet so often neglected by more traditional talent strategies. In this book, I have extrapolated the logic of playing to strengths into exploring how we tap into the rich and diverse talent of all humanity.

Call to action

The closing chapter is a call to action, suggesting we need to rewrite the DNA of our organisations and beyond into all the governance systems our world depends on. It was during the writing of this chapter of DPO that I began to sense the initial inspiration to write *Designing the Purposeful World*.

Activity 4.1: PrimeFocus and the purposeful world

We have been applying the principles of *Designing the Purposeful Organization* as depicted in the PrimeFocus framework for over thirty years and have seen similarity in how the principles play out at the levels of personal, team, department, organisation and global corporation. Before reading further, how do you think this insight might be extrapolated to the design of a purposeful world? Make your notes below:

Chapter reflection

Score the following statements out of ten where:

0 = not at all
2 = a little
4 = moderately
6 = mainly
8 = significantly
10 = completely

1 I have a good understanding of how leaders "create conditions" for success in any context.	
2 I appreciate how the thinking described in this chapter might help us to bring leadership to global challenges that may otherwise seem too big to deal with.	
3 I can describe my personal purpose in the field of sustainability.	
4 I have a vision for the world that is consistent with the SDGs.	
5 I already engage with others on sustainability and intend to do more.	
6 My life is supported by good systems, processes and other structures that enable me to contribute to the SDGs.	
7 I understand my personal values and how these fit with the SDG agenda.	
8 I am measuring my contribution to the SDG agenda.	
9 I know what success means to me regarding the SDGs and how this makes me feel.	
10 I know which of my personal talents are most valuable in playing my part for the SDGs.	

Comment

By the way, don't worry if some of your answers to the above questions are less positive than you may wish. Hopefully, by the time you've read this book to the end and participated in further exercises, your position will have changed.

Chapter 5

Several fractal leaps of scale

From single cells to humanity

It has been proven that life will adapt according to its context. Identical stem cells placed in Petri dishes containing different substances can be grown into heart, liver or brain cells. In other words, their purpose will evolve in response to their context. When cells come together in order to survive, a new consciousness arises at the collective level – that of the human being (for example). In turn, the purpose of the human will evolve in accordance with its context. Similarly, humans will come together in organisations and establish a collective purpose at the corporate level. This "zooming out" and finding self-similar patterns is the principle behind fractals as defined by Benoit Mandelbrot. Therefore, it is suggested that we can take the principles behind *Designing the Purposeful Organization* and extrapolate them to the design of a purposeful world.

Introducing fractal mathematics

Benoit Mandelbrot (1924–2010) is famous for his work in the field of fractal mathematics. He is credited with coining the word *fractal* and for explaining the mathematics surrounding the "roughness" of objects, such as clouds and coastlines and the associated principles of "self-similarity". Mandelbrot's work explains why, in many things, patterns seen at one level are similarly repeated when the observer either zooms in or out during their observation. We see this clearly in nature as we view the leaves of ferns and in physics when we compare the way electrons move around the nucleus of an atom to the way planets move around their sun. The more we come to understand fractals, the more we can find answers to seemingly big problems, such as how organisations work and, pertinent to this book, how to design a sustainable world.

Activity 5.1: Fractal mathematics

Take time out to read or watch more about the wonders of fractals and Mandelbrot's work by searching the Internet, especially Wikipedia and YouTube. Also search for fractal images to see some amazing fractal patterns. In what way might a grasp of fractal mathematics be helpful in determining the conditions in which a purposeful world might manifest? Make your notes in response to this question below.

The ability to zoom in and out and recognise patterns

I have written about fractals in this chapter following my summary of *Designing the Purposeful Organization* because of a firm belief that the principles I wrote about in that book are entirely applicable to designing the purposeful world. For around twenty years now, I have been helping organisations and their leaders to envisage the future they want and to create the conditions in which that future will manifest. Those eight "conditions for success" were discussed in the previous chapter: purpose, vision, engagement, structure, character, results, success and talent. They will be explored further in the chapters which follow.

When I first discovered fractals, I quickly grasped that the application of these eight conditions worked perfectly at different levels within an organisation. So, for example, a team purpose and vision must fit within that of the wider organisation. The same principles adopted by a team to engage its members can be used in similar ways to engage the whole organisation. The structures that are apparent at the team level also apply to the organisation as do the values and behaviours that define the team's character. The same applies to the way we measure results, establish a sense of success and make use of the talents brought into play by those involved.

Grasping these principles in the context of an organisation is relatively straightforward. We can zoom in from the corporate perspective to see similarities with the individual and their career and further out to see what happens in the case of a global corporation comprising many subsidiary businesses.

Zooming in – lessons from cellular biology

Fractals become even more interesting when we zoom in or out further still. Doing so, for me, simply confirms my suspicion regarding the natural order of life and how indeed to design a purposeful world. But let's zoom in first.

Scientist and author Bruce Lipton writes and speaks extensively about the "wisdom of the cells" that make up all living beings, including our human bodies. The proof is overwhelming that cells will actually change their behaviour if their environment changes. When he was a stem cell biologist, Lipton learnt that, by placing them in different fluids in Petri dishes, stem cells can be changed into the building blocks for different human tissues and organs. This technology is being used in modern transplant medicine.

What I have gleaned from reading several of Bruce Lipton's texts is that individual cells each have a purpose and one that will shift in some respects as the cell's environment shifts. I say "in some respects" because there are some aspects of the cell's purpose that remain constant. At the most fundamental level, the cell seems to want to thrive alongside its neighbours. Its natural stance is to co-operate with other cells, to the extent of dividing duties and working progressively as a community. Unsurprisingly, this is exactly what happens for us as human beings.

Our bodies are made of some 70 trillion cells, each of which is a living being in its own right. What is particularly interesting is that, at this mega-community level, we somehow experience consciousness as if we were just one entity. So, for most of us, our dominant consciousness is at the "me" level. As humans, most of us feel pain and take action mainly as individuals. I say "most of us" because there are exceptions.

Zooming out – the ability to feel at the collective level

As our mindset matures, we begin to feel and be more conscious at a collective level. There are many examples of people who genuinely put others' wellbeing ahead of their own most of the time. It's as if they experience the feelings of others even more than their own.

Activity 5.2: From me to we

Think of a few different examples of people who genuinely put the wellbeing of others before themselves. To what extent would you say they are experiencing consciousness at a "we" level? Make your notes below:

You may have come up with some very different types of examples. Perhaps you thought of people who have been famous for their selflessness such as Mother Teresa, Nelson Mandela, Mahatma Gandhi or Martin Luther King Jr. Or maybe the players in your local football team, your parents or the organisation you work for. People seem to lose their sense of self when they are placed in a context with others that they care deeply about. Like Lipton's stem cells, they seem to change in their nature and take on functional roles that are in service of a greater whole.

When this happens, we actually begin to have feelings or, to put that another way, become conscious at the "we" level. This can play out as empathy, where people truly feel another's feelings or indeed the pain or joy of the world. Many faiths speak of the "one-ness" of humanity or even of life as a whole. James Lovelock coined the term Gaia meaning Mother Earth and encouraged us to think of our world as a single organism that wonderfully regulates its own systems in balance.

It's very easy to see how exploring the fractal nature of consciousness can become a highly spiritual pursuit. Personally I find this to be inspiring and it provides fertile ground for profound self-development. But I would also encourage the deployment of fractal logic in numerous practical situations and ways.

Specifically, as we shall explore in the next eight chapters of this book, I invite you to adopt a fractal enquiry for each of the eight "conditions for success" outlined in *Designing the Purposeful Organization*. One approach I adopt is to consider each of the eight conditions first at the level it is easiest to grasp, perhaps the level most familiar. Just as an example, let's consider a team, maybe a team we are a part of.

The power of collective purpose

So, in terms of purpose, we can ask why the team exists. Maybe it is to fulfil a particular purpose for the wider organisation. We can test and hone this purpose by examining it from the perspectives of different stakeholders: the team members, those who supply services or resources to the team, those in receipt of services, those who fund the team and perhaps the wider community. We can ask how the evolving sense of purpose can be used to motivate team members, maybe by celebrating achievement. As we explore what makes team purpose, we gain clarity on the very nature of purpose itself and gain insights into how purpose can be worked with at a global level.

By understanding fractals we realise that what makes purpose work in a team is very similar in the world-wide perspective. At this level, we know to take account of different stakeholders: the different people of the world, those who have plenty and those who do not; people of different faiths or beliefs; and we must also remember that we humans are not the only stakeholders in the system. At a global level we have suppliers to humanity in the form of other life forms and even the very energy we receive from the sun. We too are suppliers to the system who might be regarded as our customers. We care for animals and plants and we neglect our relationship with them at our peril. At this macro level we simply must be sensitive to the whole system in order to thrive as a life form alongside others.

Sharing a vision for humanity

Considering our vision, many of us will know the excitement when our favourite team envisions winning the league at the end of the season and how we must celebrate each step along that path.

The thing that really inspired me when I read the draft of the SDGs was that this was a vision for the world. It was exactly what I had dreamt of for most of my life. Furthermore, as I have engaged with thousands of others since that first reading, I have discovered that most people share a similar vision. So, in a fractal manner, we can imagine how we would use the vision of winning the league with our team and adopt a similar stance with the world. We can celebrate milestones, we can train for the next stage, we can share the dream and excite others.

Engaging the whole world

The third condition in DPO is that of engagement. If our team never engaged with players, fans and others about winning the league, the chances of success would be slim.

So it is with the SDGs. We need to engage with all of humanity in as many ways as we can, through our families, schools, communities, workplaces and governments. We need to use as many media as we have available to us: conversations around the dinner table, in bars, at conferences, on social media, on the news and in newspapers.

Complex but powerful structures

DPO advocates structure to support the delivery of any purpose and vision. Our team will have a structure with people playing in positions to make the effort effective and systematic.

Organisations have also developed many sophisticated structures to support their teams in this way. They have clear charts showing how people report and relate to each other. They may have weekly team meetings to keep people informed and aligned, noticeboards to share information, team building sessions to keep morale high, document management systems to enable easy access to valuable data. They have plans, targets and milestones.

At a global level the same logic applies. However, there is an important aspect that is vital here and fractals help us to understand it. Even though this book is about designing the purposeful world, the structures required for such an endeavour play out at every level. The United Nations provides amazing structures and processes to maintain efficiency at that level. But we may not be playing at that level. We may decide that we can be of greatest service in the context of our organisation or community. We may describe this as our calling. We may decide to play at multiple levels. In which case we will be wise to ask what structures, systems and processes can support us at these various levels.

To give a few personal examples, my company has adopted the Global Compact as a "system" for engagement in the wider world of work. I have formed a United Nations Association (UNA) branch in my home town to engage other like minds in our area. I have set up a Facebook page for SDGs to celebrate progress all over the world towards their delivery and I have redesigned some of my family and home systems and processes to play a greater personal role – such as switching to renewable energy and buying more local produce.

The character for global success

So it is with character. Great teams have a way of performing and behaving that gets the best out of their members both individually and collectively. Some teams will formalise this, and some just make it happen naturally.

At the global level, life gets somewhat more complex but the same principles apply in a fractal manner. Fortunately there are some amazing systems available to help us understand and tune values and behaviours across the world. My favourite are the Cultural Transformation Tools (CTT) offered by Barrett Values Centre. Founder Richard Barrett is a colleague and friend for whom I have great respect. Barrett's CEO, Phil Clothier, also a great friend, and I have had some inspiring conversations about how to work with values at a global level to increase the likelihood of delivering on the SDGs.

Measuring the delivery of a purposeful world

In DPO I describe "results" as the logical, rational measurement of our progress towards our vision. For our team, the league table and various performance metrics will help us tune the team to enable success in the league.

At a global level, the United Nations has established a comprehensive set of targets that will be measured to give us confidence and correct our strategies as appropriate. But, as discussed above, we should also consider the purpose of our personal or local projects to contribute to the SDGs.

On the SDGs Facebook page, we measure the number of people liking, following the page and engaging with posts. We notice what works and what doesn't. We now have thousands of followers and some posts have reached audiences in excess of this number through onward sharing by others. The reach of the page is growing exponentially, and I'm conscious that many others will be using social media in similar ways. Through these actions, taken by several enthusiasts, millions of people are becoming increasingly inspired.

Finally, as I have refined my home energy systems, I have been measuring my energy use and source, conscious of the effect this has on my carbon footprint and personal impact on climate change.

Sharing the success of a better world

I describe success as being both a feeling and an outcome. Where results are logical and rational, success is emotional. Many people feel a sense of success at a personal level but in DPO I describe how to create a felt sense of success in a team or organisation. In a nutshell, this happens when team members share their personal senses of success and then discuss how they are inspired by what they've heard. This sharing is a systematic joining up of feelings that results in a felt sense of success at the team level.

Fractals help us understand that a similar mechanism happens with something as huge as the SDGs. When we share what excites us about progress in delivering the goals, others get excited. They may respond with new insights that inspire us even more. This seems like the journey I have been on since 2014 with people like Al Gore inspiring me about the possibilities of climate change mitigation; Steven Pinker plotting the almost unbelievable progress we have made towards world peace despite popular opinion to the contrary; and Elon Musk with his amazing progress towards a solar-powered world of electric cars, home power storage and rooftop generation.

The talent to perform on the world stage

Finally to the last condition in the set of eight featured in DPO, that of human talent. Our sports team positions people in the team according to their natural talent. This determines the positions team members play and also how the collective gels as one unit.

If I was to ask how we point the talent of the world at the delivery of the SDGs, it might seem like too big a question to answer. But fractals give us so many clues. Daniel Coyle in his book *The Talent Code* describes how masters of any talent practice for thousands of hours to become masters, building always on their natural ability.

Laurence Boldt wrote a book that changed my life and also provides clues. In *Zen and the Art of Making a Living* he describes that success lies in answering three questions. A: What talents or resources do we have? B: What do we care about? C: What is stopping us pointing A at B? So at the simplest level, I would encourage each one of us to explore what it is we might bring to the delivery of the SDGs and whether any of the goals particularly resonate.

For me, whilst all the goals resonate, I find it very easy to relate to clean energy and climate change. Maybe it's the fact that in a "past life" I was an engineer and have been trained to think about energy systems or maybe it is a growing realisation of our dependency on a life-sustainable climate and associated ecosystem. For this reason, I particularly enjoy supporting renewable energy companies like Good Energy in the UK and working to put my own house in order by the way I purchase, generate and use energy, including in my modes of transport.

Time to dive deep

I hope the above mini discussions explain how fractal thinking can help us identify what works at any level and transpose that thinking to support SDG delivery. In the chapters that follow, we will work together to take a "deeper dive" into each of the eight conditions for success in our quest to design a purposeful world.

Chapter reflection

Score the following statements out of ten where:

0 = not at all
2 = a little
4 = moderately
6 = mainly
8 = significantly
10 = completely

1 I appreciate how "fractal thinking" helps us to bring leadership to global challenges that may otherwise seem too big to deal with.	
2 I can sense how my personal purpose is converging with that of the world as described by the SDGs.	
3 My vision for the world I'd like to pass to future generations is consistent with the SDGs.	
4 When I engage with others in the work I do, I feel equipped to relate it to the global context as described by the SDGs.	
5 I know which of the systems, processes and other structures that provide method in my life might also support delivery of the SDGs.	
6 The same personal values I live out at work and home also support the SDG agenda.	
7 I can bring the same logic to measuring my contribution to the SDG agenda as I do to measuring my performance at work or elsewhere in my life.	
8 I am ready to find others to share my feelings of success with regarding the SDGs.	
9 I know which of the talents that make a difference in my career can also be used to support delivery of the SDGs.	
10 I am inspired to discover how much more of who I really am can be tapped for SDG delivery.	

Chapter 6

The purpose of humanity

This really is the "meaning of life" chapter – "Why are we here?" We explore a range of thoughts on this and extrapolate what has been discussed in the previous chapter. The hypothesis for testing is that all of life exists to thrive and celebrate itself – collectively. If the purpose of life is indeed to "thrive in community and celebrate the process", then it would seem reasonable that this should be the mission for humanity. This also carries the responsibility for sustainability so that life can continue to fulfil its purpose long after we are gone.

Introduction

If, as I propose in *Designing the Purposeful Organization*, vision is the manifestation of purpose at a particular time horizon, then the SDGs really are a powerful vision for humanity. The time horizon is 2030 and so far I have asked thousands of people to close their eyes and imagine the world at that point. They always describe a world consistent with the SDGs, even if their particular focus is on some goals more than others.

So, if this is the manifestation of a purpose for humanity or indeed for life, then what is that purpose? It is useful to know this as it explains the raison d'être for the SDGs and, thinking longer term, it will help us to establish a vision beyond 2030 when it becomes appropriate to do so. The other reason for understanding our purpose is that purpose is a useful tool for delivering our vision, as we'll see shortly.

The eternal question

Since humans first set foot on our planet, my guess is they have been asking and thinking about the purpose or meaning of life. Like many people, I too ask the question and have frequent conversations with others who are willing to engage on this topic.

Purpose as the pursuit of human needs

There are clues regarding the purpose of life, especially human life, when we examine what matters to people. In 1943, in his paper "A Theory of

Motivation", Abraham Maslow proposed that we humans seek to fulfil a series of needs, starting with our most basic "survival" needs such as food and water. With these needs met, we seek safety from harm and then to live in "social" relationship with others. According to Maslow, we then begin to grow a sense of self-esteem by feeling that we are of value in society until, ultimately, we become all that we can be and our self is "actualised". This, now famous, "hierarchy of needs" is shown in Figure 6.1 and gives us a real sense of the evolution of our sense of purpose.

This evolution of purpose hints at a progression from fear for our demise to a love of life and meaning alongside others.

In more recent years, Richard Barrett expanded on the thinking of Maslow and especially on what matters during self-actualisation. We see more than a hint of Maslow in Barrett's "seven levels of consciousness" which we will explore in Chapter 10 of this book.

Barrett suggests that what matters to us, or our "values", can be diagnosed and plotted to provide a clear picture of where our attention is and an indication therefore of our purpose. We shall explore Barrett's work in Chapter 10 when we examine values and their importance in designing the purposeful world. But for now, let's just acknowledge that our needs and our values will greatly influence how our purpose in life is felt.

In search of one simple, inspiring purpose

So, it seems that our purpose, that of humanity as a whole or indeed all of life, is complex. It will depend on who is looking at it and what their context

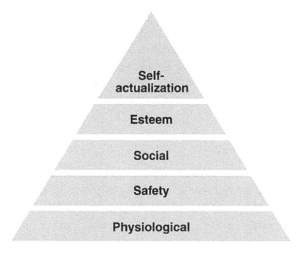

Figure 6.1 Maslow's hierarchy of needs

is. Nevertheless, wouldn't it be wonderful if we could find a way of expressing this purpose in such simple terms that we can all buy into and which embraces all variations from the most basic needs suggested by Maslow to the most ambitious contributions anyone might make? I have tried many ways of expressing this and the answer I keep coming back to is that the purpose of all of life is this:

to thrive in community and celebrate life itself

It is true that in the most threatening of contexts (at the base of Maslow's hierarchy of needs), where just surviving as an individual is all we can think of, the idea of thinking of or collaborating with others or celebrating life might seem beyond our care. Then again, if life didn't hold some promise of joy, if it didn't give us something to celebrate with others, then why would we even bother trying to stay alive.

I don't claim to have defined a meaning of life that will work for everyone at any time and it will always be a meditation but it works for me and it seems to resonate with those I have shared it with. I also like this definition as it seems to work not just for humans but for all of life.

Activity 6.1: The meaning (or purpose) of life

It isn't my choice of words or how I've put them together that really matters. It's the essence that counts. Of course, people have been trying to understand the meaning of life for thousands of years, so I don't suppose this activity is a new one. But, as an anchor to understanding the whole of this book, why not craft your own purpose for life in a few words and write it in the space below. Feel free also to contact me and let me know. Or post it on the SDGs Facebook page.

It seems to work for the simplest of life forms

My logic for this is that there seems to be plenty of evidence that all living organisms seek to thrive. As previously noted, stem cell biologist and author Bruce Lipton taught that identical stem cells could be placed in different contexts in Petri dishes and would adapt in order to thrive, thus creating a variety of tissue types that could ultimately be used for medical purposes. At every scale of magnitude we see the same pattern. Insects, birds, animals, humans and communities organise themselves with the intention of thriving, consciously or otherwise.

Life thriving alongside life

I have also proposed that it is essential that life thrives alongside life, in community. Again, this is complex, especially as we see life taking life in order to survive and thrive. This seems to be part of the grand design.

Not only does life need to co-exist with life, it also appears to be designed to thrive in community. Look at the amazing way the 70 trillion cells in our bodies have congregated and self-organised to form the life form we call "me". In the same way, you and I have come to identify along with 7 billion others to form the "we" we know as "humanity". And in the same way that I am conscious at the "me" level, many of us are increasingly becoming collectively conscious at the "we" level.

Consciously thriving at the global level

Of course, there is much more to life than humanity. James Lovelock described our interconnected global organism as Mother Earth or Gaia. This mega-community of a myriad of life forms suggests at the "energetic" level a joined-up global consciousness with a collective intelligent awareness. At a biological and material level it is a mega ecosystem that (on a good day) seems to work.

Does consciousness diminish at the micro level as it matures at the macro?

It is interesting to ponder that when consciousness arises at the collective level, so the consciousness at the individual level seems to subside. We can all think of people who have put their families, their team, their nation or the utmost collective good before their own wellbeing. We experience it when we are with someone who is hurting. We feel their pain more than our own sensations in that time and place. We experience it when we see the harrowing images of conflict on our television screens.

We notice it when individuals are prepared to sacrifice their own life for the collective good. We might naturally think of this mainly with humans in mind but might it be possible to stretch our imagination to make sense of how life is sacrificed for life in the food chains of our global ecosystem?

Behaviours shift as our consciousness grows

As we associate ourselves more with the whole, amazing things begin to happen. For example, Otto Scharmer, author of *Theory U: Leading from the Future as It Emerges*, describes how our listening changes as we evolve. At the most primitive level we have the closed mind that simply listens for facts to support our own stance. At the next level we listen with an "open mind" in the knowledge that others have valid data we can learn from. As we mature, we learn to listen with an "open heart" where we have empathy for the feelings of another. Finally, in our most evolved state we become less aware of our separateness and more in tune with a collective future that is seeking to emerge. This deep listening for an emerging future is what Scharmer refers to as "generative listening".

Living together as one, literally

This is exactly the nature of a collective purpose. It has a life of its own. Frederic Laloux, in *Reinventing Organizations*, describes purpose in highly evolved "teal" organisations as being independent from any one person's control. The CEO and other executives are in service of the purpose (as opposed to owning it) as are all other parties associated with the organisation.

Why does all this matter?

The beauty of having a simple purpose for all of life is that it provides an anchor for all we do. It provides alignment and commonality in a world that is all too often divided. Seeking to thrive together and wonder at life itself is surely something we can all buy into without feeling we have to leave our particular faith or philosophy behind. A simple purpose for life has the power to unite and build bridges.

It is therefore critical to understand how the context of a human being might affect how it will respond to a "thrive and celebrate the process" hypothesis. Clearly someone living in a context of love and sufficiency might find thriving and happiness easier than someone living amidst starvation and fear of violence.

Maybe, as suggested by Maslow in his "hierarchy" of needs, there is a basic need to survive first before someone has any chance to rise to the purpose of thriving and celebration. However, this doesn't seem to be as straightforward as it seems. History abounds with stories of heroes and heroines who have laid down their lives through love for others. We also know of others who have more than enough to survive and yet live in constant fear of losing even a small proportion of it.

This conundrum is at the heart of whether humanity will indeed deliver the SDGs. With a collective heart of love we could do this today. In a world of fear, we have no chance to succeed. Phil Clothier of Barrett Values Centre describes his number one value as "love". Indeed, he also describes this as his purpose. Love might easily be defined as the desire to share life with others, to thrive and be happy. In which case, there will be those amongst us who might understandably simplify my thoughts on the purpose of life still further as being "to love".

Love does not prevail uncontested

Interestingly, as I write this chapter, much of the world seems to be at a fear versus love crossroads. Western politics are more polarised in this manner than has been the case for many recent years. We saw this playing out with Brexit in the UK and in the US and French presidential elections. Whilst this has been a sad and frustrating distraction from our commitments to sustainability, my optimism believes that this may be a necessary playing out of the human psyche. Those at the "love" end of this spectrum might consider those at the other to be cruel, discriminatory and selfish. In turn, they themselves might typically be regarded by their opposites as soft, trendy do-gooders, living on a cloud of delusion far removed from the harsh realities of life.

Full spectrum consciousness

The truth is that we need people at both ends of the spectrum to "generatively" listen to each other so that, if nothing else, a better level of understanding is achieved. Such wisdom, as Richard Barrett might say, leads to the development of "full spectrum consciousness" where people truly have values across all levels of consciousness. People with full spectrum consciousness still respect and value safety whilst taking action to make a difference in an holistic context.

Politics of fear and love

What is interesting is the trend of the division between the politics of fear versus those of love. Whilst not a strict rule and without wishing to polarise or alienate, there is evidence to suggest that, in general, the nationalistic, isolationist politics of fear find more favour with older people and those who are less well educated. The younger generation, who are often also better educated, show a distinct preference at the ballot box and in their life choices for fewer social barriers and greater compassion for others.

The rise of the politics of love

The point of raising this topic is not to make this text political. It is simply this, that as older people die and education becomes prolific and widespread, we might speculate that the politics of fear will diminish and the politics of love will prevail. Humanity may then find it easier to focus on its true purpose. Such progress would be well facilitated by a focus on the SDGs and an ultimate trend towards the politics of love may give us more reasons for hope – that might easily have been added to that message in Chapter 2.

The fight back of fear

However, this won't happen overnight and it won't happen without some initial discomfort. Make no mistake, those living in fear and promoting fear in defence of their fear-born self-interests will not lie down easily. As the world naturally becomes more joined-up and compassionate, those who are against such trends may well resort to desperate measures. We've already seen the lengths they will go to in order to further their cause, including the adoption of aggressive tactics and blatant use of misinformation or "false-facts".

Nor is it the case that everyone who is materially wealthy will find safety and security in their wealth. There are plenty of such people who have what to others may seem a disproportionate fear of losing even a small proportion of a vast wealth and a need to grow their wealth far beyond their needs. Similarly there are many materially poor people who would give their last cent to help their neighbours and who leave their doors wide open to strangers.

The need for compassion

Wealth and power in the hands of people who lack compassion is a dangerous thing, especially when they find allegiance amongst those who have little, with whom their words of fear resonate. There is little point in compassionate liberals ignoring this as their high ideals will simply serve as a turn-off. Instead, they need to listen generatively to the needs of the many and find a way of responding that makes more sense than the alternative messages of fear and defence.

Wealth and poverty, fear and love it seems are largely an attitude of mind. But research into the topic of "happiness" suggests that it is to be found in the spirit of love, generosity, compassion and abundance.

Educated, loving people come to the conclusion that "we're all in this together". They know that big topics like climate preservation and world peace require a strong collective effort. They know that inequality adds fuel to fear and causes division. Ultimately there is no winner from the path of fear. But the path of love must embrace the needs of the many in order to succeed.

The power of celebration

We need to find a way to make sustainability attractive. In this respect and in our quest to accelerate our journey to the purposeful world, a key additive to our fuel is the power of celebration. Celebration of purposeful successes strengthens the purpose itself. I have noticed, by the number of views we get on our SDGs Facebook page, that when we celebrate progress on the delivery of the goals, people are drawn into curiosity and find inspiration. They become less interested when we over-focus on the problems and more interested when they see solutions. This is the case when we post new innovations in renewable energy and when we celebrate victories for gender equality such as women piloting planes in countries where they're not even allowed to drive. It is the case when young people find business opportunities cleaning our oceans and when permaculture solutions return Africans to ways of farming that were lost during colonial diversions from their heritage. It is the case when the life of an endangered animal is saved and when young women defy their oppressors to receive the education they deserve.

Activity 6.2: Celebrate what's right with the world

Too often the voice of sustainability is one of doom and gloom. This is perfectly understandable and we do need to acknowledge the problems with compassion and sincerity. However, if we are to succeed in preserving our planet we must find ways to make doing so a truly attractive proposition. And it is. Science and innovation is sexy. Progress is delightful. Education is exciting. Children playing safely is a joy to behold. Preserving and witnessing life on land and under the sea is amazing.

National Geographic photographer Dewitt Jones reminds us to "celebrate what's right with the world" and I would encourage every reader of this book to watch Dewitt's video of the same title. There are plenty of tasters and trailers of this work on YouTube if you'd like to sample his work before getting a copy of the full version. Watch as much as you are able and inspired to do and make your own notes below on what you think is "right with the world" and how this can be leveraged in the interests of sustainability.

Every time we celebrate progress relating to a purpose we inspire more of the same and the power of the purpose grows. In turn, a powerful purpose will drive a compelling vision which is my cue to the next chapter.

Chapter reflection

Score the following statements out of ten where:

0 = not at all
2 = a little
4 = moderately
6 = mainly
8 = significantly
10 = completely

1 I have given some serious thought to what the purpose of life might be, especially since finding out about the SDGs.	
2 I have written down my own version of this purpose,	
3 I have shared this with others in conversation.	
4 I have shared thoughts on the purpose of humanity with others on social media.	
5 I can see clearly how the SDGs are a vibrant playing out of the very purpose of life.	
6 I have considered how these insights on the purpose of life will have a bearing on how I live my life.	
7 I have reflected on the love/fear spectrum and worked out where I stand.	
8 I have compassion for those living lives that are fear based and thought about how I might engage with them in a way that is neither patronising nor antagonistic.	
9 I choose to be more purposeful going forward.	
10 I have made a commitment to do at least one thing to make my life more purposeful.	

Chapter 7

A shared vision

How the SDGs define our future in a way we can all accept

If our purpose is to thrive in community and celebrate life itself, we need to know where this takes us. Vision is simply the manifestation of purpose at an important time horizon. In several workshop sessions that I've run with a variety of audiences (schools, universities and professional groups), people have been asked to close their eyes and envisage a better world at a 2030 time horizon. On every occasion the world they describe is totally consistent with the SDGs, conveniently as described and adopted by world leaders at the United Nations in New York in September 2015. This is especially convenient because we therefore don't have to sell the SDGs as a future to "buy into". It is a future that the vast majority of humanity already holds dear in their hearts. Of course, as with any vision, people see some aspects of this better future more clearly than others. As discussed earlier, this is a function of who they are and their context. Nevertheless, there are few disagreements regarding any significant aspects of the bigger picture.

We're all different

As discussed in the previous chapter, our purpose and the purpose of all of life is arguably to *thrive in community and celebrate life itself.* This is such a vast purpose and will naturally play out in a myriad of ways for different people and even for different life forms. I am immediately reminded of the way an eagle celebrates by soaring high in the sky, a lion basks in the African sunshine and a flower blooms in joyful colour. It seems that the more we look, the more we see life in celebration with life.

Vision – the manifestation of purpose at a particular time horizon

A vision is a playing out of purpose at a particular time horizon in a particular context, in a manner that can truly be visualised in advance. The SDGs are time-stamped at the year 2030 and they paint a comprehensive picture, not just of humanity, but of our whole planet, including ecosystems on land and in the sea and about our precious climate on which all of life depends. The other goals focus on humanity, on how we co-exist, in harmony, with compassion and in collaboration.

The wonder of the SDGs

It is amazing that world leaders have actually signed on to this vision. Or is it? The fact that I have engaged with thousands of people, young and old, and discovered that they too share the same vision, is inspiring to say the least. That said, we should bear in mind that, at the time of writing, most have never even heard of the SDGs.

Discovering that people everywhere, with only few exceptions, share this common vision is something to really build on, something to unite us all. The work of drawing this vision from people, without lecturing them about the SDGs, is both important and urgent. For it is this work that will inspire people, not only to see the same future, but to find a way to be part of it.

Space to engage

When I work with groups of people on this topic, I'm very careful to create the space in which their vision for 2030 can emerge. Before the SDGs are even mentioned, I ask them to close their eyes and go on a journey to the world they'd like to leave to other generations in 2030, visiting two or three different parts of the world. They get to choose where they go. I merely suggest they move around in their mind's eye and look at what's happening in cities and in the countryside. They notice what people are doing, how they relate to each other and what's going on around them. Only when they have a clear picture and have had the chance to share with and hear from others do I ask if anyone knows what happened in New York in September 2015.

When people realise that the world they long for is the same world that leaders have agreed to deliver, they become enthused. The reason this happens is that their vision has been made legitimate. Hearing that people around them also share the vision makes further dialogue not only possible but highly probable.

The same vision through many lenses

Let me be clear, having said that we all share the same vision, we don't all see it in quite the same way. This is so with any vision. When I ask people what they see, they relate aspects that are relevant to them. This is because we all live in different contexts and have different stories to tell. Some things matter more to some people than to others. But there is little or no conflict. Just because one person wants to see an end to poverty, it doesn't put them at odds with someone who wants to take action to mitigate climate change. In fact, there is so much overlap across the SDGs that many of the things people feel inspired to do will hit several goals.

I have a friend in Malawi who has an embryonic business installing solar panels in rural villages. Clearly this is an SDG 7 activity but it definitely hits SDG 1 (less poverty); SDG 3 (better health through removing alternative fuel sources that pollute the airspace in people's homes); SDG 4 (education by enabling children and others to study by LED lighting and charge their phones and tablets); SDG 11 (sustainable cities); SDG 13 (climate) and probably most of the other goals too in one way or another.

My friend clearly has a vision for clean and modern energy supply. He didn't even know about the SDGs but now that he does, it has given legitimacy and extra power to his vision. He now truly feels part of the greatest adventure that humanity has ever conceived.

Accelerating progress towards the vision

One of the most powerful ways to accelerate progress to the goals is to provide space in which visions can grow. At our local UNA branch in Harrogate, we run a process based on Action Learning Sets (made popular by Welshman Reg Revans whose father used the method to investigate the sinking of the Titanic) where people can share the challenges they seek to tackle based on the SDGs. The other attendees offer in their own most valuable questions and comments thus enabling the challenge owner to make new informed commitments to progress. This process not only makes the vision clearer for the person who brought the challenge but it also repeatedly causes new visions to emerge for many of the attendees.

Activity 7.1: The vision becomes clearer

On most journeys, as we approach our destination, the place we are heading towards comes into view; perhaps the mist lifts and we get more excited about what we see. Turn back to the notes you made at the start of this book regarding the vision you have for 2030. Is there anything you'd like to add to this picture, perhaps another layer of detail? Or perhaps you'd like to pick another time horizon that is particularly meaningful to you. Perhaps you are about to embark on a new career or, like me, you are contemplating your so-called retirement. What do you see at this time horizon?

Taking a single step toward the vision

As your vision becomes clearer, it will probably also become more compelling. As this happens, it may be time to take action. Don't worry that you don't yet have a comprehensive plan. Ask what small step you could take today that will take you closer to your vision. Even if it's just to spend time with someone else who you can talk your ideas through with. Or perhaps it's a note to undertake further research.

One of the key factors that I believe will drive progress is the fact that it is now so easy to conduct research online. As my vision becomes clearer, I go online to read or watch documentaries on what the future might hold for many of the goals. I have discovered that there are countless enthusiasts and experts who have recorded interviews and conferences available at the click of a button. So when I have moments of doubt on questions such as "Is world peace even possible?", I discover that, among the sceptics, there are people like Stephen Pinker who not only believe it but who have collected compelling data that, despite occasional setbacks, it is actually happening.

Moreover, many of these people are doing amazing things to contribute to the possibility. This type of online research makes our own vision more clear. With clarity comes hope and confidence, and with confidence comes conviction and action. So just imagine, if even 1% of humanity begins such a journey, how the world will turn!

Visions within visions and alongside visions

In Chapter 6, I introduced the concept of fractal mathematics and the thought that the subject is invaluable in the design and creation of a purposeful world. Fractals really do help us to understand how the power of vision works.

When I facilitate workshops in organisations, I usually ask one or more of the senior leaders to share something about the future of that organisation with their wider leadership team. I encourage them to preface their vision with some words and stories to affirm the purpose of the organisation and to focus their thoughts on those aspects of the future that will be the most inspiring and also those that will be the most challenging. I also ask them to acknowledge that they don't have the whole picture and that the reason we are gathered in the workshop setting is to co-create the future together.

My colleagues and I then run tried and tested processes where we ask each participant to envisage the future at the given time horizon as they see it. Then they pool these insights with those of their colleagues. They are normally reassured and delighted that their insights are not dissimilar to those of others and that there is very little disagreement amongst them. The fact that they all see the future in similar but complementary ways affirms the need for diversity and teamwork in making this future a reality.

So it is with the direct reports of the senior team. They may not see the big picture with such clarity but they will probably see their piece of it in greater detail. It's not that their vision is any less important, it's just that it reflects their context.

This idea of visions within visions and complementary visions is, I believe, one playing out of Mandelbrot's theories of fractal mathematics.

This is a useful insight, to say the least, when we consider the design and delivery of a purposeful world and the role of the SDGs in such a quest.

I have proven beyond any shadow of a doubt, in my mind anyway, that the SDGs are the most powerful vision for the world, ever. I have proven that some (fractal) part of this vision is held dear in just about everyone's heart. I have seen at workshops that different people will see progress to different goals differently and with varying degrees of enthusiasm. I have listened to Bill Gates talk about the need for large-scale genetically modified crops to feed the world and the lesser known Luwayo Biswick in Malawi speak just as passionately about the power of small-scale permaculture with intimate respect for natural ecosystems to bring abundant harvests capable of feeding whole villages in what is frequently ranked the poorest nation on the planet. And, whilst I have my preferences, I also keep an open mind to all possibilities that might render images of starving children and adults to the history books.

The more we look the more we see

Curiosity is a key factor in providing clarity of vision. It seems that, for every problem we face and every aspiration we have, there is someone who has either been there before us and holds the answer or who has done something similar that will help provide the answer.

This was the case when I got curious about climate change and loss of species. I chose to ask unthinkable questions such as, "Is climate change reversible?" and "Can we recover species that have been lost from the ecosystem?"

I discovered that, for climate change, reforestation actually has the potential to absorb carbon dioxide from the atmosphere, thus reversing the effects of processes that release the same gas. I discovered that humanity is investing in some amazing projects, such as the "green wall" which is being formed by growing new forests and vegetation along the edges of the Sahara Desert. This caused me to imagine a world where the land-cover of forests could increase and with it the possibility, not just of climate mitigation, but also of more abundant life on the surface of the earth. It made me wonder what might happen if, as well as reducing carbon dioxide emissions, we asked those responsible to create forests with the capacity to absorb twice the quantity emitted. Might the audacity to ask unreasonable questions and develop previously unimaginable visions prompt someone else to build on the idea and do something. If writing this book prompts no other action, my efforts will not have been in vain.

On the subject of species loss, I discovered that there are people seriously working in the field of "de-extinction", using DNA samples from the scarce or extinct to reintroduce to other similar live species. I only discovered this when I was researching what could be done about species loss. It is a controversial topic, with some suggesting it is playing God with genetics. However, an Internet search brings countless pages of information on the topic which assured me this will be part of our future and could help us undo the damage we have done through our lack of care for species in the past. The point is this, curiosity develops vision and vision inspires action.

"One small step"

Those famous words uttered at the time of our first walk on the moon are so true in the field of any endeavour. Just as curiosity drives vision, so does any action. In my workshops some participants comment that the challenges they see and the visions they hold are so vast, they don't know where to start and they feel unable to make a difference.

It is absolutely my experience that no action is too small to take. This is not a task we face alone. We are working on the biggest team task known to humanity. If the steps we take head in the right direction, we usually end up closer to our destination. We arrive at a new vantage point from where the vision is much clearer. Occasionally we might take a wrong step but even then the learning will inform a more productive change of direction. We will meet obstacles but the strength of our purpose and vision will find a way round, over or through them.

Walking with others

Have you ever noticed how much easier and enjoyable a journey can be when we walk or travel alongside others. Finding other people who have similar interests and enthusiasm to our own is a great way to make sure we keep the pace of our journey or perhaps even accelerate progress to our goals and, at the same time, the SDGs.

Keeping the vision in our mind

When we have a powerful vision it begins to infect the way we live our lives. We notice things that relate to our vision, whether it's in the news or the places we visit. I also find that this works in reverse. I find I can deliberately heighten my sense of vision by deliberately managing my news feeds, by consciously visiting inspiring places and by engaging with relevant art visually or through music.

Songs from my past like John Lennon's "Imagine" or Bob Marley's "Redemption Song" now resonate with new meaning for me. I've grown fond of the "Playing for Change" movement, performed magically by musicians playing together from all around the world without leaving their neighbourhoods (you'll find them on YouTube). I often use their songs as people are arriving at my workshops. I have a favourite playlist called "Inspiration" where each song reminds me in some way of the better world I hold in my mind's eye.

Activity 7.2: Keeping the vision alive

There must be thousands of ways people around the world keep their vision alive. Perhaps you could note your favourite ways below or share them with me and others on the SDGs Facebook page.

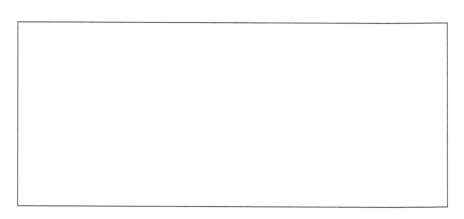

The main thing is we become engaged in a fabulous journey. This feels like a cue for the next chapter.

Chapter reflection

Score the following statements out of ten where:

0 = not at all
2 = a little
4 = moderately
6 = mainly
8 = significantly
10 = completely

1 My vision for the world in 2030 fits well with that described by the SDGs. 2 I have identified the goals that resonate most for me. 3 For these goals I can describe how I'd personally like the world to be. 4 I have identified personal goals to take me closer to this vision. 5 I have already taken some small steps towards my goals. 6 I have shared my thoughts with others. 7 I have researched what others are doing to take the world closer to my vision. 8 I have shared their progress verbally and through social media. 9 I keep my mind open to the best efforts of others even when they're not the same as mine. 10 I keep my vision alive through places I visit, the way I manage my personal space and the art I enjoy.	

Chapter 8

Engaging absolutely everyone in the SDGs

Without engagement and consequent action, a vision is just a dream. When we engage, people become inspired, they pull together and change happens. It is not sufficient to leave the delivery of the SDGs to world leaders – and the world is already working towards the goals, whatever our leaders do or say. However, the more we engage people of all ages, the greater our progress will be. In this chapter I reference some of the great work already being done through leaders, through the work of the United Nations, through organisations in all sectors, through communities and through the efforts of individuals. Everyone gets to participate in the greatest team event humanity has ever staged.

Start with "Why?"

Let's begin this chapter with some more fractal thinking linking right back to our purpose:

to thrive in community and celebrate life itself

It is such a natural phenomenon for life to gather with life to support this idea of thriving in community. As Bruce Lipton reminds us, the very cells of our body do this amazingly. They form super-teams with other cells to create our organs. We see them wiring up communications and transportation systems so that information and resources get to where they are needed to enable them, and consequently us, to thrive.

How can I work with you?

Just by scanning the world we live in we can see this engagement happening in an infinite number of ways. It is as if life at every level is enquiring "How can I work with you to help us all to thrive?" Then, at the end of a job well done, we rest peacefully or celebrate enthusiastically. Think of the joy we get by being with others or by being in nature, watching young animals playing or bees pollinating flowers. The way our ecosystems work is truly miraculous.

It's as if engagement is so natural that we don't even have to think of it. However, the purpose of this chapter is simply to speed up the engagement process by promoting some thoughts on the many ways we can encourage and support each other to choose ways to benefit the world that fit with our unique personalities and contexts.

Sharing the vision

I am reminded of the story of the blind Indians who approached an elephant and used their hands to "see" what it was like. The one at the tail described it as like rope. The one grasping the tusk described it as hard and rock-like. The one at the trunk said it was rather like a big snake. Whilst the one poised near a leg disagreed with them all, asserting that it was much more like a tree.

I began to hint at the need for engagement in the previous chapter when I spoke about sharing our vision with others and facilitating opportunities for them to do the same. I spoke of the work undertaken by our UNA branch in Harrogate. I've also previously alluded to the workshops I run in schools and universities where young people can consider their vision for 2030 and make the connection to the SDGs in the context of their own education or career journeys.

But it doesn't stop there. There are limitless ways you or those around you could engage further with the SDG agenda. Here are just a few for starters:

1   Ask friends, family and others you meet what they would like to see for the future. Create an initial space for discussion.
2   Consider running a workshop where you can systematically explore people's visions and share news about the SDGs. There are plenty of resources in this book that will support you.
3   If facilitating a workshop isn't what you do, consider setting one up and inviting a facilitator to work with you. If you don't know anyone, feel free to contact me for suggestions.
4   Be expansive with the range of groups you could engage: schools, universities, employers, local authorities, community groups, professional institutes, conferences, retirees, charities. You'll be amazed how receptive people are on this topic.
5   As you gain confidence in engaging others, train them to do the same. I was delighted when the young leaders at AIESEC International asked me to train 400 of their international community at the Youth Action Summit at the UN in New York. This was within just three months of the goals being agreed by world leaders in the same building. See the case study on this later in this chapter.

6 Consider running a project that engages others in the SDGs. Maybe it's to serve meals for the homeless in your town or to fit solar panels to the roof of the local school. Remember, one thing leads to another.

7 Social media is a great way of engaging others on the SDGs. There are lots of platforms you could use for this. You may wish to join our community of thousands of SDG advocates around the world at our Facebook page (www.facebook.com/SDG2030).

8 If you'd really like to support the SDG delivery from a UN-recognised perspective, why not consider joining a Branch of the United Nations Association (UNA). If there isn't one near where you live or work, consider starting your own branch as I did. It is very straightforward and your National UNA office will guide you through the process.

9 If you're with a company, you can engage through membership of the UN Global Compact. Joining the Compact involves making a public declaration that your organisation signs on to and promotes the principles of human rights and environmental sustainability. You can add your declaration to your company website and make specific reference to the SDGs as my colleagues did at Primeast in the UK.

10 Perhaps you'd like to engage with your local or national politicians, as I did at the start of 2015 prior to the UK general election. I asked four of our local candidates if I could interview them on an undisclosed topic that would determine how I would vote. Two of the four agreed and their interviews are still on YouTube. I voted for the one that gave the most inspiring and convincing commentary on delivering the SDGs.

Activity 8.1: Sharing the vision

Take a few moments to reflect. Who do you feel called to engage with at this stage of your journey? And how might you best go about doing this? Perhaps you can highlight the list above and make notes below.

Engaging for self-motivation

The points above are principally about you engaging with others to spread the word and enthuse others in the SDGs. But there is another form of engagement which is about meeting others to inspire your own deeper involvement.

Since embarking on this journey, I have taken every opportunity to meet with others who are sharing the path from many varied perspectives. I have spoken with solar and other renewable energy companies, town planners, people working in education, people who run infant homes in the poorest parts of the world and farmers seeking to make permaculture available to villages in Africa. Every time I get curious and arrange to meet to explore their world, I am inspired by their enthusiasm and dedication.

This conscious following of my own curiosity and sharing the journey with others has been a great stimulation for making things happen in my life and the lives of others around me. Let me give you a personal example of just one engagement which relates to SDG 7 (clean energy).

Case study 8.1: Meeting Chris at Leeds Solar

You may recall from an earlier chapter that I began my working life as a student engineer with the Yorkshire Electricity Board in 1972. I then spent thirteen years as an electrical engineer working on the mains distribution system: building substations, laying underground cables and erecting overhead lines. I organised teams (of mainly men) to construct and maintain the system and to respond to electrical faults at any time of the day or night. I loved the satisfaction of finding the line that had been brought down by heavy snow and fixing the system so the farmer could have the supply back on in time for milking. The best reward was a cooked breakfast in the farmhouse with the linesmen as a thank you for working all night in the snow.

No surprise then that SDG 7 and clean energy attracts me. I know the difference that solar panels, modern battery technology and SMART metering can and will make to our environment and climate change (SDG 13).

My curiosity prompted me to find out who was involved in solar technology in my home county of Yorkshire in the UK. I discovered Leeds Solar and arranged to meet their technical director, Chris Platt. I asked Chris about the solar business and the challenges brought about by the government reduction to the Feed-in Tariff. FIT determines the remuneration paid to domestic and other generators of electricity from renewable energy.

Chris explained that the reduction of FIT had severely impacted the UK solar market, causing lots of solar companies to go out of business with associated loss of jobs for installers. However, he also added that the business case for solar remained compelling as the price of equipment had been falling. The net effect was that those players who had entered the market for short term gain had "packed up and gone", leaving those with a real sense of purpose to continue to make things work.

Chris agreed to join us at one of our UNA meetings and told the story of Leeds Solar. In return he received lots of great advice and support from the group to strengthen his business.

One thing leads to another and, having met Chris and heard his story, I asked him if he'd like to survey our house in Harrogate, explaining that we were undecided whether it would be viable due to the possibility of a house move in the coming years.

Chris conducted the survey personally and offered a range of solutions with the business case for each one. We decided to fit a bank of fourteen solar panels on the south-facing roof of our house and to support this system with a modern battery unit. We now have a system that generates most of our electricity demand during daylight hours and stores most of what we don't consume in batteries for use in the evening when the sun has gone down.

It's early days yet, but the system seems to make us almost energy-neutral as far as electricity is concerned.

In terms of engagement, I'm naturally very proud of our installation and enthuse about it with friends, neighbours and colleagues, some of whom are keen to "have a look".

Curiosity prompts action prompts curiosity

The whole world over, people are becoming increasingly curious about what they could do to play their small part in delivering the SDGs. Many will be curious and do nothing, but some will take action and their action will prompt curiosity in others and so the change accelerates.

For me, it didn't stop with the installation of solar panels. I also swapped my gas-guzzling Harley-Davidson motorbike for an electric push bike whose very small battery is now charged by solar energy. Whilst my biker chums might think I've gone mad, my sustainability conscience simply got the better of me. I could no longer ride around on a two wheeled monster that consumed more petrol than most people's cars.

I've further committed to buy no more petrol or diesel vehicles. And the energy I do buy comes from Good Energy, a renewable energy supplier in the UK, who (by the way) also buys our surplus energy and pays our Feed-in Tariff income.

Big hairy audacious goals

I first learnt about "B-hags" from a colleague in Dubai who used "big hairy audacious goals" as a symbolism for ambitions that summed up the strategic direction almost to the point of being unbelievable. It's a sort of "how big do you dare dream" concept.

Whilst my "big hairy audacious goal" for solar isn't going to resolve global warming single-handedly, it does serve to keep me inspired and heading in the right direction. One day I aim to own an electric campervan, fitted with solar awnings and travel the world on sunshine only. Flight of fancy? We'll see. My engagement with others on this mad idea is moving me closer and closer to this far-fetched dream becoming a reality. I've already discovered a company that makes solar awnings for buildings and in my mind's eye I can see them attached to either side of my electric van parked up in a campsite charging the batteries for the next stage of my journey.

My second SDG 7 B-hag is the development of an eco-fisherman's cottage in the Northumberland coast in northeast England. This one has already moved closer to fruition during the course of writing this book. As part of our pension investment, rather than buy an annuity, my wife Frances and I decided to buy an old fisherman's cottage in the beautiful coastal village of Seahouses. Such dwellings were typically heated by coal and a "back-boiler" and their stone walls uninsulated. We have replaced the heating system with an air-source heat pump that takes heat from the atmosphere outside the cottage and pumps it around the interior. The next stage is to add solar panels to the roof to provide the equivalent electricity as that required to drive the heat pump and our other energy needs. Our intention is to make the cottage available as a holiday-let so that visitors can experience clean energy in action and hopefully be inspired to do something similar.

Activity 8.2: Elon Musk

If you're interested in SDG 7 and would enjoy large scale inspiration on the prospects for solar energy, battery technology and electric vehicles, check out (on YouTube) Elon Musk, founder and CEO of Tesla and his amazing cars, sexy solar roof tiles and "Powerwall" home battery solutions. Make your notes on how this inspires you below. Alternatively, make notes below on what might be an inspiring B-hag for you.

Chris at Leeds Solar is not the only person our small UNA group in Harrogate has engaged with. We have had several people in our "hot seat" on a Monday evening at the Primeast office in Harrogate. Each one of them has made commitments to accelerate their work in delivering one or more of the goals. Here are a few more examples at headline level only:

Jane: had a passion for SDG 4 on education. She also had a passion for wildlife (SDG 15). She committed to supporting a school in Zambia with links to a national park and the preservation of the rhino population. She travelled to Africa and has been working on the project ever since.

Kit: runs a business that converts classic vehicles (mainly VW camper vans) to electricity (SDG 7) and organises holidays where people can hire them to have amazing eco-experiences. He got some great ideas from the UNA members to market what he does and accelerate the growth of his business.

Mike: was concerned about how men fit into the gender equality equation (SDG 5) and brought his thoughts to the group. He has adapted his whole life to leading men's work to benefit society.

Viv: had a passion for sustainable communities (SDG 11). She and her friend Rich brought their ideas and insights to the group and became inspired to lead an even more minimalistic lifestyle. They're now working and living their dream in Portugal.

Anthony (a speaker and journalist on sustainability) and Phil (a lawyer): brought their collective thoughts on clean energy to the group and inspired everyone to do more to deliver SDG 7. Their workshop at UNA led to the invitation for Tom to attend (see below). Anthony's podcast "The Sustainable Futures Report" is also now a regular feature on the SDGs Facebook page.

Tom: is CEO of a City Council and shared the impressive actions being taken already to deliver clean-air solutions for the city centre. He committed to run events to encourage businesses and others to make the city carbon-neutral.

Lindsay: works with people approaching or in retirement. She was passionate about how the SDGs could inspire this group of people to lead engaged and meaningful lives beyond their full-time jobs.

Matthew: began funding and building schools in Tanzania and has thus provided education to hundreds of children, almost single-handedly.

Chris: is a trustee for St George's Crypt in Leeds, UK, who helps many of the city's homeless people struggling with the challenges of poverty and often ill-health.

Claire: is a fund-raiser for the Open Arms Infant Home in Malawi. She is based in Harrogate and sought the support of our UNA group to determine how best to inspire supporters for this wonderful organisation.

All the people described above have benefited from having a few like-minded people with diverse skills and experience around them, supporting their SDG journeys.

Engagement is leadership

We may not immediately make the connection, but it is important to remember that engaging with others on topics we believe in is the most powerful form of leadership. It is authentic and compelling and will resonate significantly with many of the people we meet. This is especially true when our enthusiasm spills over into action that helps to deliver the SDGs. For me personally, I can think of no better role than being an energetic advocate for the most significant commitment humanity has ever made.

Activity 8.3: The tango and the SDGs

Continuing the leadership theme, I invite you to search YouTube for the following TEDx video, "what the tango can teach about leadership". Hear what Sue Cox has to say about leaders, followers, collaboration and the dance. Then, having read this book this far, make notes on where the responsibility for leading on the SDGs lies and where followership fits in. And where are you? Are you leading, following, in the dance or simply watching from the edge of the dance floor? With these insights, what will you be inspired to do differently?

Case study 8.2: Stretching the imagination – The Venus Project

Some of the ideas on engagement discussed above can be regarded as "extracurricular". They may simply get "bolted on" to our lives as optional extras. Others, like Chris at Leeds Solar (see above) become integrated into people's careers. They are all steps in the right direction of positive change. However, it seems to me that it is worth stretching our imaginations still further if we are to achieve transformation.

Take the very first goal, that of eliminating poverty. We can achieve a great deal through many acts of kindness and sharing. But poverty doesn't go away. To truly eliminate poverty we need to change the norms of the system whereby people engage with each other across the world.

One truly fantastic image of what this might look like is Jacque Fresco's The Venus Project. It is a vision of what humanity might look like if we all engaged according to our talents in support of a planetary purpose. Just like the cells in our bodies, in The Venus Project, we contribute according to who we are and what the system needs. And just like in our bodies, the communication and resource systems are aligned to the common good rather than to channel wealth and power to a privileged few.

Activity 8.4: Inspiration from The Venus Project

The Venus Project is featured prominently on the Internet so why not take a look at one of the many videos and make notes below on how the work of Jacque Fresco inspires you.

There are more people and institutions than we might think who are actively working on such advancements, ranging from social entrepreneurs like Fresco to governments like that of Finland.

Case study 8.3: Finland and a guaranteed income

As reported in *The Guardian* newspaper at the start of 2017, Finland became

> the first country in Europe to pay its unemployed citizens an unconditional monthly sum, in a social experiment that will be watched around the world amid gathering interest in the idea of a universal basic income.

Under the two-year, nationwide pilot scheme, which began on 1 January, 2,000 unemployed Finns aged 25 to 58 will receive a guaranteed sum of €560 (£475). The income will replace their existing social benefits and will be paid even if they find work.

The reason I quote these social advancements here is that, alongside these radical developments, there seems to be a more subtle social shift in attitude that makes such development even thinkable.

This is what I mean when I also talk about shifting engagement from the extracurricular to the curricular and from the curricular to the transformational. Or, to put it another way, from "bolt-on" to "business as usual" to "business far from usual".

Case study 8.4: Kerr Mackie primary school in Leeds, UK

Helen Crowther is deputy head at this school in Leeds, not far from Harrogate and actually the city where I was born and raised. She heard about the workshops I was running on the SDGs for children and asked if I could do one for about 200 children aged between 7 and 11 at her school. I did this in 2016 and caught up again with Helen in June 2017 to see what had happened since.

Helen and the children at the school had become very interested in the SDGs and had made several connections with other initiatives and taken some great actions.

They have an RRSA (Right Respecting Schools Award) Steering Group focused on the UN Rights of the Child. Here are just a few examples of their initiatives:

- The children talked about making links between the SDG workshop and the school curriculum.
- They hired a "pop-up shop" and created displays on Life below Water (SDG 14) and Life on Land (SDG 15). They shared this learning with parents and the public.
- They appointed "energy monitors" to keep an eye on a range of savings and improvements that could be made, including energy, water, recycling and fair trade.
- The have made links between life on land and the climate and have supported National Clean Air Day. Children in year 6 made posters that were used in the campaign.
- They've had a "Walk to School Week".
- The children wrote to Prime Minister Theresa May and the Home Office to draw attention to the government's commitment to take 3,000 refugees but only actually taking 350.

- They have explored SDG 16 and enquired what more could be done to secure peace in our world. They have linked with a local university to get theology students to engage with the children on this topic.

All in all, the children really enjoy being part of something that is truly a global agenda as "global citizens", and the parents have been impressed by the impact this one topic (the SDGs) has made on the quality of learning.

Helen had this to say about the work:

> Our pupils are compassionate and are keen to get involved and make a difference. Using the SDGs as a platform for this, gives their ideas and actions gravitas and a purpose. For children younger than 11 to inspire their families, community and peers is impressive. We are so proud of our pupils and their concern for our world.

Activity 8.5: Engaging others according to our passion

Reflect on the above case studies. What is it that you feel particularly passionate about regarding the SDGs? And how might you step out in a way that might engage and inspire others? Make your notes below.

The power of social media

Social media is a phenomenon that has crept up rapidly on humanity, virtually unannounced and with immense power to engage widely for change. I find it hard to remember life before we began using this modality for sharing our lives, views and news. Yet, these days many people get most of their news and learning via social media, even if it is simply that our friends and associates draw our attention to mainstream news, through their social media posts.

When the SDGs were announced, I thought I'd try using social media to share information and progress on the goals. I set up the Facebook page at www.facebook.com/SDG2030 and began posting articles on the SDGs, usually introducing them with a short title naming the relevant goal or goals. At first I invited a couple of hundred of my personal friends and contacts but through their engagement in this feed, the numbers involved have grown to several thousand and continue to grow exponentially. I expect our followers to be in five figures in 2018 and many more by the time we reach 2030. We now also have a small team of editors who join me in making posts to the site. If you really like this idea, you could even contact me to become one of them.

Activity 8.6: The power of social media

Feel free to join us at the SDGs Facebook page and to consider how you might use the power of social media to engage others in the most important topic ever. Make your notes below and let me know your thoughts and ideas for further development.

Chapter reflection

Score the following statements out of ten where:

0 = not at all
2 = a little
4 = moderately
6 = mainly
8 = significantly
10 = completely

1 The concept of engagement as a natural phenomenon of life has inspired me to be more involved in the SDGs. 2 I shall engage more with my friends and family to encourage the delivery of the SDGs. 3 I shall engage with leaders and others at my workplace to make the connection between the work we do and our contribution to delivering the SDGs. 4 I shall encourage my place of work to consider joining the UN Global Compact. 5 I shall personally join my local branch of UNA or other organisation to get more involved with the SDGs. 6 I shall engage in my local community to raise awareness and involvement with the SDGs. 7 I shall consider undertaking a personal project that contributes to the SDGs and sharing my involvement with people I know. 8 I shall engage with my MP to encourage a greater prominence of the SDGs in their manifesto. 9 I shall enquire with my local authority regarding their alignment to the SDGs. 10 I shall engage with news feeds and generally on social media to raise awareness of my friends and contacts to the SDGs.	

Chapter 9

Building the structures for change

As a consequence of engagement, people become involved in all manner of activities. This chapter examines how structures can be put in place to make change more effective and efficient. We identify the role of organisations, networks, social media, plans, legislation, funding mechanisms and so on. The topic is massive so this narrative is really only the tip of the iceberg. However, we also revisit the nature of fractals and some of the latest thinking regarding how, when inspired by a powerful purpose and vision, individuals can respond quickly to build structure around their efforts to make a difference.

Systems and structures for life

Life has always strived to put in place structures that enable it to thrive. Think of the 70 trillion or so cells that live in community in our bodies. These organisms, each alive in its own right, have figured out how to congregate in communities and in physical structures such as our vital organs. They share resources such as nutrients in our blood stream and communicate via a complex nervous system that works most of the time without us even thinking about it.

As humanity, we too have worked out how to live in community. We have put in place transportation systems to ship resources around our world. We have a complex nervous system in the form of the Internet and various telephone networks. Most of the time for most of us, the structures of our world just work without us being overly conscious about it. That's the nature of structure, it guides the way we do things and often dictates how effective we are. Too much or wrong structure complicates life, as we see with the bureaucracy that characterises many of our organisations and parts of civil society. Too little structure and we have anarchy and nothing seems to get done.

Structure to support what we do

Of course, behind the scenes, each one of us is "doing our bit". According to our personal and shared aspirations, we either use the structures that already exist or we put in place new ones.

As people engage around the SDGs, structures begin to emerge for the system to work and improve. The most obvious structures are those of governments and of the United Nations and its subsidiary organisations. These are the principle vehicles for change to happen. But it doesn't stop there and nor should it. More will probably happen outside the obvious formal structures of international government than within them. We will all find ways of putting structure around our own efforts to deliver the goals. By structure, I don't just mean organisations; I also mean processes, plans, communications media, anything tangible which helps to make the SDGs delivery more likely or more efficient.

The goals themselves provide a convenient structure giving clarity around seventeen ambitions for the future. Each goal has a number of targets and associated measures which are easily accessible online for all to see. We'll say more about this in Chapter 11.

At a personal and family level, I like to brainstorm the things I could be doing to advance sustainability using the seventeen goals as a structure to focus my thinking. So if I have a few spare minutes waiting for my train to arrive, I will often pick a goal or a few goals and ask what I already do to support the delivery of that goal. I then ask what more I could be doing and make a new commitment. I have discovered that this practice has been personally challenging in a positive sense and has really improved my behaviours. I have discovered once again that small steps lead to bigger steps.

Case study 9.1: UNA Harrogate – a structured approach to SDG engagement

In our local community, I decided that I could provide a convenient structure for SDG advancement by establishing a local branch of the United Nations Association. My initial intention was to find a local branch and simply join it. But the nearest branch to my home town in Harrogate was about seventy miles away in Kingston upon Hull. So I decided we should have our own and I set out to make it happen. This is all about adding structure that will support the delivery of the goals. In my view, there should be a UNA branch in every town and city around the world. So if you're looking to provide much needed structure to make the world a better place, there's an easy win for you.

Even the way we structure our UNA meetings in Harrogate is important. Each time we meet, we have a volunteer in the "hot seat" who describes their progress towards a particular goal in about fifteen minutes and who asks the group to help them make new, more progressive commitments. From this point, we use a standard "Action Learning Set" method (yet another well-tested structure) so that everyone in attendance offers in their most powerful questions and comments to the "challenge holder" until that person knows exactly what they need to do next. At this point we ask them to make a suitable commitment to action. In this way our meetings have been structured deliberately to produce action rather than just being a talking-shop.

From macro to micro – structure all the way

Can you see how the above UNA example demonstrates how simple structures taken at the macro are translated through other structures right down to personal commitments that make a real difference? Following the path from macro to micro goes something like this:

1   The big picture is the need to evolve humanity and the planet we call "home" so it can be enjoyed by future generations of all life forms.
2   World leaders at the UN (a structure for the global evolution of peaceful and sustainable government) have translated this big purpose to a vision comprising seventeen compelling goals (the SDGs). Note that they used the added structure of a UN working group to do the hard work of drafting the goals in a manner that would secure agreement.
3   The UN in turn has the support of yet another structure in the UNA as a mechanism for ordinary people like us to engage in our communities to challenge and advocate for its work.
4   Within the UNA I was able to form a group specifically focused on the SDGs and was able to draw on yet another structure (the long-established method of Action Learning Sets) to make our meetings compelling and action oriented.
5   We also use the structure of a meetings calendar, an email distribution list and the Internet to organise our meetings.
6   Finally we also draw on the support of Primeast to provide excellent meeting space (a further predictable structure) for our UNA work.

Case study 9.2: Polly Higgins – strengthening legislation as essential structure

Global agreements such as the SDGs provide a structure that, to some extent, holds leaders to account, allowing scope for sanctions if they stray from the agreed path. And the United Nations, with its various agencies, works tirelessly on subsets of the sustainability agenda.

However, there are people who feel that the voluntary nature of agreements like the SDGs is insufficient. People like Polly Higgins believe that change will be effective only when there is clear legislation to enable it. Think about it, good legislation is a very effective structure for making sure things happen (or don't happen) in a way that is conducive to the overall good of society.

Polly Higgins describes herself as the lead advocate for ecocide law. According to her website at PollyHiggins.com she is "rethinking law from one simple overriding principle: "First do no harm." A crime of ecocide is a practical law: the State has to prosecute on your behalf (i.e. you do not have to sue), your rights are protected and the government has to stop the harm. It's simple. Destroying the earth is a missing crime (not yet on the statute books) exacerbated by State failure to assist those who are at risk of climate displacement.

Activity 9.1: Ecocide action

Take a look at Polly's website and maybe watch one or more of her videos. Make a note of your thoughts about the need for legislation on ecocide versus a more voluntary approach in the box below. Is this the sort of structure the world needs to ensure appropriate action?

Strengthening corporate structures

Think too about the plethora of other organisations that take the SDGs and align to them in the work they do. I have had several conversations in the corporate world with business leaders highlighting the many advantages of being part of such a powerful global agenda. They also know that being seen in this way inspires all their stakeholder groups, which is a powerful performance enhancer.

In national, regional and local government, the goals provide a convenient and logical way of organising the work they do. I'm pretty sure that, as we move closer to 2030, we'll see more explicit links between government and the SDGs. After all, the goals seem to represent a manifesto at every level that everyone wants to be delivered.

Distributed leadership and how we provide structure for the SDGs

My colleagues who specialise in leadership development have invested significant time and attention examining how the nature of leadership is evolving and therefore how leadership development must also evolve. In this chapter

I want to highlight the main thrust of this thinking and say why I think it's important with regard to how we provide appropriate structure to support delivery of the SDGs.

We live in a world that many commentators describe as "VUCA" (volatile, uncertain, complex and ambiguous). This demands a style of leadership that is mature and transformational, characterised by being able to see an emerging future, collaborate with others who might be very different in character, and create conditions that enable the organisation to serve this future in a relevant and appropriate manner. By the way, we also describe these conditions as: purpose, vision, engagement, structure, character, results, success and talent. You will recognise these as the components of PrimeFocus, the core chapters of this book and of DPO.

To enable this to happen, leaders need to be developed in four ways:

1   to develop skills (in their tool box) that enable them to perform (often referred to as "horizontal development");
2   to develop a mature mindset where they understand they don't have all the answers and must collaborate with others in order to find a way forward (often referred to as "vertical development");
3   to achieve alignment of purpose that makes sure that all parts of their organisation are working to the same end and that the organisation is aligned to the wider purpose of humanity; and
4   to achieve behavioural alignment so that their behaviour contributes positively to a wider corporate culture which in turn is conducive to delivering the organisation's purpose and vision

The net effect of all this thinking on leadership is that leaders let go of the reins of their organisation and allow leadership to be distributed to everyone involved. Instead of *owning and driving* the purpose, they become *servants of the purpose.*

This is vitally important when we come to think about structure. In progressive organisations structures become flatter, people are empowered and there is less hierarchy and autocracy. People are freed up to contribute the best of who they are in service of a purpose they believe in. These attributes characterise what Frederic Laloux calls "teal organisations" in his book *Reinventing Organizations.* I mention this elsewhere in this book because distributed leadership is exactly what we need on a global basis to deliver the SDGs. I therefore encourage you to read more or watch Laloux on YouTube. His thinking is progressive.

Think about the alternatives. If we wait to be told what to do in support of the SDGs, their delivery may never happen. If we leave everything to the UN or world leaders and expect a top-down system to emerge that tackles the lot, we will be disappointed.

My hypothesis and belief in distributed leadership plays out as follows. The SDGs provide a very clear vision for humanity. In the subtitle to this book,

I describe it as "a blueprint for humanity". In this book, for those who haven't made the connection for themselves, I have asserted that, as with every vision, the goals are a "playing out" of purpose – the purpose of life itself. So the point is this, we only need to look into our hearts and collaborate with others to work out what we need to do. We should seize any convenient structure that makes it efficient to step into action and remove any hierarchical structures that restrict empowerment and personal progress.

Activity 9.2: Exploiting distributed leadership

How does the concept of distributed leadership help your involvement in delivering the SDGs? Do you feel empowered to take action? A key feature of distributed leadership is knowing when to consult with others before taking action. Who would be a good person to talk with to test your ideas for action and perhaps help you put some appropriate structure around your plans? Make your notes below.

An infinite array of structure to take advantage of

One of the difficulties we may find as we start to act is navigating the world's systems to deliver sustainability. There are so many organisations that have set up to make one contribution or another and there are lots of systems and processes that we might adopt. In this respect, I would urge that before adding to this complexity, you might think about using existing "best structure" before adding to the complexity.

The structures that could support the delivery of the SDGs are as many and varied as your imagination.

Think for example about:

- the Internet as a system that provides connection of thoughts and ideas;
- within this, the specific opportunities associated with websites and social media;

- National Parks as places to protect ecology and showcase best practice;
- ecosystems themselves as natural structures that support the interdependency of life;
- families as opportunities to engage on sustainability across generations;
- schools and universities where people can learn about these matters and reach out further into the community to engage parents and local people; and
- within places of education, the curriculum and how this might be tuned to the SDGs.

The list is infinite.

Activity 9.3: Structures to support our inspiration

Look back at your earlier notes on which goals inspired you and in what way. Make notes below on any structures, processes and systems that will help you contribute to the delivery of the goals that inspire you most. You may wish to brainstorm some ideas with a group of friends or colleagues and then evaluate your list to identify some items worth adopting.

Case study 9.3: The UN Global Compact

When I was looking at what I could do to encourage alignment to the SDGs, I consulted with my colleagues regarding our own work at Primeast. We decided that the most appropriate structure we could become part of was the UN Global Compact. This organisation provides the opportunity for organisations to make a public commitment to the following ten principles which preceded the SDGs but which also naturally encompass them.

The UN Global Compact asks companies to embrace, support and enact, within their sphere of influence, a set of core values in the four areas of human rights, labour standards, the environment and anti-corruption.

The ten principles listed under the four areas are as follows

Human rights:

1 Businesses should support and respect the protection of internationally proclaimed human rights; and
2 make sure that they are not complicit in human rights abuses.

Labour:

3 Businesses should uphold the freedom of association and the effective recognition of the right to collective bargaining;
4 the elimination of all forms of forced and compulsory labour;
5 the effective abolition of child labour; and
6 the elimination of discrimination in respect of employment and occupation.

Environment:

7 Businesses should support a precautionary approach to environmental changes;
8 undertake initiatives to promote greater environmental responsibility; and
9 encourage the development and diffusion of environmentally friendly technologies.

Anti-corruption:

10 Businesses should work against corruption in all its forms, including extortion and bribery.

My colleagues chose to join the Global Compact, make the associated public statement and begin a programme of continuous improvement in line with the SDGs. You can explore the Global Compact online and discover other organisations like ours that have signed up to its principles. You can read their letters of commitment and follow their progress.

Chapter reflection

Score the following statements out of ten where:

0 = not at all
2 = a little
4 = moderately
6 = mainly
8 = significantly
10 = completely

1	I understand how systems, processes, plans and physical spaces for engagement all provide structures that can be enhanced to make the delivery of the SDGs more effective.	
2	I have efficient structures in my life to support my involvement in the delivery of the SDGs.	
3	I have plans to enhance my structures for SDG delivery.	
4	My local community is well-served with structures that will support the delivery of the SDGs.	
5	I have thought how I might engage with local structures such as institutions or local government in support of the SDGs.	
6	I have other plans to enhance local structures for the SDGs.	
7	I understand why legislation might be useful for making SDGs a reality and will find out more about legislation and crimes such as ecocide.	
8	I can see the value in adopting distributed leadership as a vehicle for the SDGs and will play my part as a leader alongside others to deliver positive change for future generations.	
9	I will lobby government and international agencies in support of the SDGs.	
10	I understand that structures do not work without the corresponding will and behaviours to support them (and I'm therefore keen to read the next chapter)!	

Chapter 10

Why our values and behaviours (our consciousness) must evolve to support change

Richard Barrett has written extensively on the subjects of values, behaviours and culture. His organisation, the Barrett Values Centre, has produced some powerful tools to take stock of the values in any system: organisation, community, team, individual, nation or even humanity. In this chapter we ask to what extent are the values of these entities likely to support the delivery of the SDGs and, importantly, what can be done about it. We stress the need to be methodical about cultural change at all levels and provide links to methods we know will work. Incidentally, please note that the PrimeFocus model introduced earlier uses the word "character" instead of culture, reflecting the fact that when values are played out at the personal level, the word "culture" isn't commonly used.

Case study 10.1: Culture change in the UK electricity industry

In the 1990s, I was involved in some pretty major change programmes as the UK electricity industry moved from public to private sector. At one point I was programme manager for an initiative called the Competitive Metering Programme (CMP) for Yorkshire Electricity, one of the then principal electricity supply and distribution companies. We had twenty-one projects in our programme and were very proud of the progress we were making. We collaborated with two other electricity companies in the development of new software solutions that would allow electricity meters to be read by different companies from those that supplied and distributed the power.

Like so many organisations in technical industries, we were very good at sorting out all the technical, logical and rational aspects of change. We had great plans, systems and processes. We were clinical in our documentation and resolution of issues, risks and changes and we had some great minds working on the computer and communications systems and associated software.

At one point, due to the national importance of these changes, our programme was audited by external consultants. We passed with flying colours on all aspects of change except one. That was the measurement and management of the culture associated with the change. The consultants pointed out that the culture required to run the new privatised and competitive industry we were forming would be very different to the one we had all known during our nationalised era.

This was all new territory for me. I knew very little about culture and was totally unaware that it was something that could be measured and managed systematically. I consulted with those who knew more about the topic than anyone in our team and, in the end, it was agreed that I should travel to Detroit in the US and become accredited as a practitioner with an organisation called Human Synergistics, which specialised in cultural change and leadership development.

This was a first step in me becoming increasingly interested in leadership and organisational change – and ultimately in the global change associated with sustainability.

Measuring human values in a system

Since becoming accredited with Human Synergistics, I have explored other methods that support cultural change. One that impressed me greatly was that offered by the Barrett Values Centre (BVC). Richard Barrett developed his system for values measurement in a similar fashion to Abraham Maslow who described human needs in the form of a hierarchy as shown in Chapter 6 where we explored the relationship between human needs and the purpose of life.

Barrett extrapolated this thinking and translated it into the language of values. He provided a catalogue of human values which he allocated to his "seven levels of consciousness" framework as shown in Figure 10.1. Note the similarity between the bottom half of Barrett's model and Maslow's hierarchy of needs.

The Barrett method is now well-established globally and has been used to diagnose the values of individuals, teams, organisations and even nations. If you're interested, BVC offers a free Personal Values Assessment or PVA at its website which will gives you a full computer-generated report on your personal values together with implications and suggestions for personal development. It is a great initial step to sampling the methodology before using it on a larger scale, such as a team, organisation, community or nation.

Values and sustainable development

What has all this got to do with the SDGs? Well everything actually. As I learned in the 1990s, change programmes can fail miserably if we don't consider the cultural aspects of change. And the SDGs represent perhaps the most ambitious change programme ever signed off by world leaders.

Seven Levels of Consciousness

Human Needs Human Motivations

Human Needs	Human Motivations	
Spiritual	Service	7
	Making a Difference	6
	Internal Cohesion	5
Mental	Transformation	4
Emotional	Self-Esteem	3
	Relationship	2
Physical	Survival	1

Figure 10.1 Barrett's seven levels of consciousness

Source: with kind permission of Barrett Values Centre

When I'm working with an organisation on a change programme, I strongly advocate for the use of the Barrett method. Using a Company Values Assessment or CVA, I recommend that the organisation seeks data on three aspects:

1   What are the values brought to the organisation by those involved in the change? Ideally I recommend that this includes all stakeholder groups, such as customers, suppliers and staff.
2   What are the values that these people witness being lived out in the organisation right now? This data represents the current culture or "the way things are done around here".
3   What are the values these people would like to see in support of the particular change the organisation is setting out to achieve? This would be the target culture.

This information is invaluable and can be "cut" to find out how it varies between stakeholder groups or any other relevant criteria, such as geography or role-type.

Having the data helps those involved to see at a glance the required culture and the transition that needs to be made. It also helps leaders to understand the behavioural change that they need to make to be better able to "show the way".

This type of cultural change programme can be a very positive experience for those involved. The target culture is nearly always a more positive scenario than that experienced currently. And it often more closely resembles the values brought to the change by individuals, meaning that they can authentically be part of the solution.

National Values Assessments (NVAs)

In considering the SDGs, I encourage everyone involved to take a look at the BVC National Values Assessment Resource Guide produced at the time of the 6th National Values Coalition Meeting 26–27 September 2016 in Toronto, Canada. The data in this report show unquestionably (in my mind) that people care deeply about the topics reflected in the SDGs.

Values that appear frequently in the desired futures of many nations include sustainability, environment, community, care for the elderly, future generations and so on. By the way, for most of the countries surveyed, these values are also sadly lacking in the perceived current cultures.

The data in the National Values Surveys provide both hope and concern. Hope lies in the fact that the values held by people in these nations are likely to support the delivery of the SDGs. Hope also lies in the fact that people want to see values consistent with the SDGs in the future. The concern is that they don't see these values featuring in the way society operates or is governed right now. In a nutshell, the perceived current culture in most nations is far from positive with values such as bureaucracy, conflict, aggression, blame and poverty featuring prominently. One notable exception is that of Bhutan where the current culture is extremely positive and where values that fit well with the SDGs such as environmental protection, nature conservancy and social justice appear in the top ten.

Another factor that is playing out in all this is that what people perceive isn't always the truth. In the cultural data, people may sense factors like fear and violence and yet statistical data shows that we live in a more peaceful world than has ever existed before. Despite the obvious and terrible violence being played out in certain parts of the world, which is clearly a matter worthy of global attention, the truth is that death due to war and terrorism has been reducing for many years and is broadly at an all-time low.

Again, with notable exceptions, people are healthier, live longer and are wealthier, especially in the so-called developed world. Statistics also show that the developing world is catching up rapidly.

Should we believe what we read in the newspapers?

We might speculate that the media and those with vested interests who are practicing fear-mongering are intent on creating a world of fear because it sells advertising or generally suits personal or corporate purposes to do so. Or we might reasonably argue that it is indeed entirely appropriate that we shine the spotlight on the ills of the world in order to raise the level of urgency to sort it out.

One challenge is that of short-term versus long-term perspectives. The spotlight shone by the world media, especially the "popular press", tends to focus on the immediate and obviously urgent. Every day we witness scenes of violence and terrorism. The scenes are awful despite the long-term over-all trend of improvement. Sadly other factors don't get the attention they deserve. Climate change rarely hits the headlines and yet it is the one factor that has the capacity to "finish us all off" if we don't address it.

So what is to be done? My belief is we need strength of leadership in all quarters to move with urgency on the delivery of all the goals. Our progress needs to be systematic and on all fronts. Leadership, therefore, needs to play out in every context, personal, family, community, corporation, government, nations, international and global.

We really should be encouraged by the positive values held by the people in the world and their desire to see a better future. They need encouragement that we are, despite the news broadcasts, making progress. They need to know that every single action in support of the SDGs really matters and they need to share and celebrate such progress so that it becomes infectious and encourages others to do the same.

It seems to me that a conscious focus on values at all levels of society is going to be good for delivering the sustainability agenda. Even without the formal establishment of the SDGs, the National Values Assessments carried out by the Barrett Values Centre demonstrate that people throughout the world care about the same things that the sustainability agenda is focused on.

Therefore, the more such values assessments are undertaken and people are involved in setting the agenda for change, the more likely we are to deliver on the SDGs.

I have had several inspiring conversations with Phil Clothier at Barrett Values Centre regarding our focus when we look at the National Values Assessments. If we focus on the current cultures in countries, we might tend to despair at the negativity in the system. If, however, we look at the personal values brought by citizens to the same countries, we can see and measure the energy for change. This is, I believe, exactly the same phenomenon as when I ask any audience what world they'd like to bequeath to future generations and they describe, time and time again, a world that is totally aligned to the SDGs.

Activity 10.1: Personal Values Assessment

Search the Internet for the Barrett Values Centre PVA and complete the free assessment of your values. When you receive your personal report, make a note below of any actions you'd like to commit to in order to live your values more fully. Make a note also about which of the SDGs would be supported by simply being more of who you are.

Having completed the above exercise, consider the power of advocating a values assessment for an organisation you are part of. A Company Values Assessment (CVA) will help your organisation to be a more authentic and successful contributor to society, especially if this is done in conjunction with exploring the organisation's purpose in the wider world context.

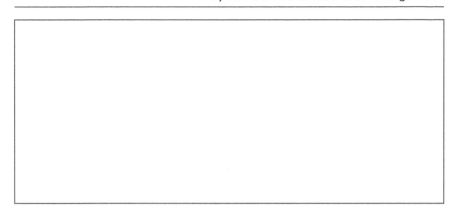

Guest contribution from Phil Clothier, CEO of Barrett Values Centre

I truly believe that the Barrett method is key to transformation in support of delivering the SDGs. As I was completing this book, I had the good fortune to catch up with Phil Clothier for one of our "SDG catch-ups". I asked him to say a few words about his take on the goals and the role of values in their delivery:

Contribution from Phil Clothier: shared values are a powerful source for personal and societal transformation

The UN Global Goals for Sustainable Development first came on my radar in early 2015. I knew from the first moment that this was an opportunity for humanity to work together to create the life conditions so that everyone on earth has the possibility to experience well-being, happiness and peace. It was also clear to me that this is a journey with the potential to liberate our core human values and bring them to conscious awareness in our hearts, minds, families, workplaces, communities and nations.

As the CEO of Barrett Values Centre, I know that I am in a very privileged position. Yes, I am very aware of the challenges and crises facing humanity and the planet but I also regularly hear stories from around the world from individuals, groups, organisations and nations where people are taking steps to make a difference and in their own special way, bringing the ambition of the Global Goals to life.

It doesn't matter what race, colour, gender we are, we are all human beings and values stand at the centre of our needs, fears and motivations for action. At every moment, whether we are conscious of our values or not, we are making decisions based on our needs and values.

We all have needs and if these are being met, we will feel positive emotions such as a sense of contentment, fulfilment and happiness. If the needs are not being met we feel negative emotions such as fear, anger, sadness or disappointment. The simple truth is that we value what we need. Needs fall into two main categories, basic needs and growth needs.

Basic needs

A basic need is something that is important to get, have or experience in order to feel physically or emotionally safe. You feel anxious or fearful when these needs are not met because they are vital for your physical and emotional well-being. Once your basic needs are fulfilled you no longer pay much attention to them.

Growth needs

A growth need is something that enables you to feel a sense of internal alignment with who you are. Once our basic needs are met we start to connect with our sense of purpose and meaning in life and find ways to share our skills, gifts and talents with others. Often a greater sense of care and compassion will arise that shifts our focus into relieving the suffering of others. When you are able to satisfy your growth needs, they don't go away; they arouse deeper levels of commitment and the joy of the soul in leading a purpose driven life.

The chart below (see Figure 10.2) shows our basic and growth needs and values mapped against the Barrett 7 Levels of Consciousness Model. It also maps the Global Goals that will help address the needs at each level.

The current reality is that the majority of the world's population still live without their basic survival needs being met and life is a constant struggle. However, just having enough money and material wealth is no guarantee of happiness and fulfilment either. Many people in the developed world live with much fear, stress, unhealthy diets and very poor quality of life.

The good news is that the world is waking up and more people are becoming aware of their values and putting them into action in daily life. As they start to make decisions based on their values, they focus on what is most important (basic and growth needs) and this makes a positive difference in their life and for the people they influence and touch at home and at work. We know that becoming conscious of our values and putting them into daily decision making makes a significant positive difference to quality of life.

In 2012, I had the realisation that we at Barrett Values Centre could make a significant contribution by creating a Personal Values Assessment. Later that year we launched this assessment and made it available to everyone on earth, free of charge. The intention is to help people explore their values and become aware of what is most important and what might be holding them back from fully living them. I have heard many stories about how this opens a doorway in people's lives to a whole new understanding of deep priorities and motivations. It only takes a few minutes to take a Personal Values Assessment and you will be sent your own report in an email. If you find it valuable, please share it with family, friends and colleagues. www.valuescentre.com/pva

Seven Levels of Personal Consciousness with the UN Global Goals

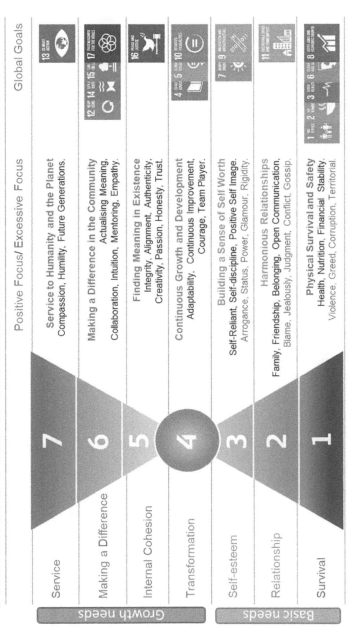

		Positive Focus/ Excessive Focus	Global Goals
Growth needs	Service — 7	**Service to Humanity and the Planet** Compassion, Humility, Future Generations.	13
	Making a Difference — 6	**Making a Difference in the Community** Actualising Meaning, Collaboration, Intuition, Mentoring, Empathy.	12 14 15 17
	Internal Cohesion — 5	**Finding Meaning in Existence** Integrity, Alignment, Authenticity, Creativity, Passion, Honesty, Trust.	16
	Transformation — 4	**Continuous Growth and Development** Adaptability, Continuous Improvement, Courage, Team Player.	4 5 10
Basic needs	Self-esteem — 3	**Building a Sense of Self Worth** Self-Reliant, Self-discipline, Positive Self Image. Arrogance, Status, Power, Glamour, Rigidity.	7 9
	Relationship — 2	**Harmonious Relationships** Family, Friendship, Belonging, Open Communication, Blame, Jealously, Judgment, Conflict, Gossip.	11
	Survival — 1	**Physical Survival and Safety** Health, Nutrition, Financial Stability. Violence, Greed, Corruption, Territorial.	1 2 3 6 8

Figure 10.2 Seven levels of consciousness and the SDGs
Source: with kind permission of Barrett Values Centre

Personal values of 405,000 people mapped against the Barrett 7 Levels

	Level 2
7: Self-less service	
	family (161078) 5
6: Making a positive difference in the world	
	humour/fun (133489) 5
	caring (123625) 2
5: Finding meaning in existence	
	respect (113005) 2
	friendship (113000) 2
4: Letting go of fears.	
The courage to develop and grow	trust (109517) 5
3: Feeling a sense of self-worth	
	enthusiasm / positive attitude (108,428) 5
2: Feeling protected and loved	
	commitment (107259) 5
	creativity (106031) 5
1: Satisfying our physical and survival needs	
	continuous learning (104588) 4

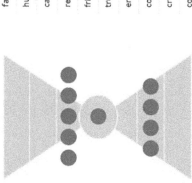

Figure 10.3 Personal values

Source: with kind permission of Barrett Values Centre

At the time this book was written, we had over 400,000 Personal Values Assessment responses from people in over one hundred nations. Here is the global top 10 personal values (in order of the number of people who picked them). (See Figure 10.3.)

This shows me a world where positive and fun loving people care deeply for each other and have a commitment to make a positive contribution to others through continuous learning and creativity.

Another reason I feel so blessed to be doing this work is that I have seen the positive impact that comes from personal, team, organisational and societal transformation. When people align around an inspiring purpose and live a shared set of values, seemingly impossible feats can be accomplished with ease. Yes it can be challenging and a little scary but there is nothing as powerful as a group of people acting with purpose and passion.

Chapter reflection

Score the following statements out of ten where:

0 = not at all
2 = a little
4 = moderately
6 = mainly
8 = significantly
10 = completely

1 I have a clear understanding of my personal values.	
2 I have made the connection between my values and the SDGs.	
3 I have studied the National Values of my country and other places I have interest in.	
4 The organisation I work with has taken stock of the values brought to work by its people.	
5 The organisation I work with has a good understanding of the values actually being played out in the workplace (as opposed to the ones posted in reception).	
6 The organisation has diagnosed the values required to deliver its intended future.	
7 The organisation I work with has considered its values in the context of an emerging and sustainable future.	
8 Leaders in my organisation have had the opportunity to leverage their personal values to deliver the aims of their organisation.	
9 They also know how their values fit alongside others in delivering global change in line with the SDGs.	
10 I have a personal plan to act authentically in line with my personal values and be a force for good in the world.	

Chapter 11

Measuring results and defining success

As any large organisation knows, we need to measure our progress towards our vision as a set of results that indicates the extent to which we are making headway as desired. The SDGs themselves are accompanied with a set of measurements that will serve this purpose for world leaders and associated institutions. There will also be those amongst us who will find roles as monitor-evaluators. These people will be able to use such data to encourage, lobby, cajole or celebrate as they see fit. But results are very left-brain, logical and rational, as indeed they need to be. Success, however, is more right-brain, more creative and inspiring. Success is both an outcome and a feeling. It is a felt-sense of progress. This can be at any level: individual, partnership, team, community, organisational, national, even humanity. The more that success is shared and celebrated, the more there is hope and inspiration. We therefore have an important role, as advocates of the SDGs, to celebrate progress and encourage more of the same. This role will be played by some more than others according to their particular sense of purpose. But no celebration is insignificant. Rather it is the cumulative effect of millions of celebrations, small and large, that will create a movement that feels successful and delivers the goals.

Reporting on global progress at the United Nations

The goals, sub-goals and specific targets relating to the SDGs are most easily accessible through the UN Sustainable Development Knowledge Platform at https://sustainabledevelopment.un.org/sdgs. These are necessarily detailed and provide the opportunity for the UN and member states to hold one another to account. They also provide an opportunity for citizens to hold their governments to account and many readers of this book may well see their personal role as being one of "monitor-evaluator" watching carefully for progress and lobbying government to do "the right thing".

Pursuant to General Assembly resolution 70/1, the Secretary-General of the United Nations, in cooperation with the United Nations system, submits the report on progress towards the Sustainable Development Goals. The first such report was issued in June 2016 and this level of reporting will repeat each year. The report is available to the public at the United Nations website.

Activity 11.1: Progress on my favourite SDG

Check the particular goals and targets for the SDG you are most interested in. Also search the Internet for progress reports at the global, national and local levels. How are we doing?

In addition, a comprehensive, in-depth report will be produced every four years to inform the high-level political forum convened under the auspices of the General Assembly. The next report will be published in 2019.

In order to strengthen the science-policy interface at the high-level political forum convened under the auspices of the Economic and Social Council, scientists who work on the report are invited to provide scientific input into the discussion, including on the theme of the forum.

The UN Secretary-General received over sixty nominations of experts in the natural and social sciences from across the world for membership in this prestigious panel. The final selection of fifteen experts was made with a view to providing balanced coverage of the various topics that could be expected to feature in the Global Sustainable Development Report (GSDR), 2019, while also respecting regional and gender balance. The GSDR will promote a holistic and science-based approach to policy measures that will advance the Sustainable Development Goals (SDGs) and Agenda 2030.

Measuring progress at national and local levels

Notwithstanding the targets and measures that will made at a global level, I'd like to reaffirm what was said earlier in this book about fractals (principles of self-similarity). Just as measures taken at a global level need to be appropriate for monitoring world progress, there also needs to be measures at the national and local levels to make sense of progress on the SDGs.

For example, here in the United Kingdom, the Office for National Statistics (ONS) is responsible for the official reporting of UK SDGs data. This includes:

- submitting UK data for SDG indicators to the UN to inform the global reporting framework (as defined by the UN Statistical Commission) and making these data equally accessible to all;
- working with official and non-official data producers to identify supplementary, UK focused indicators; and
- exploring and developing new data sources and methods to enable ONS to report data at the various levels of disaggregation.

The measurements need to be done in a thorough manner to make sure we are also complying with the wider spirit of the SDGs. Increased disaggregation of data by sex, race, religion, geography, disability, ethnicity, migrant status, age and income quintiles is critical to the "Leave No One Behind" agenda which is integral to the SDGs.

At the time of writing, when we zoom in to the local scenario, there is much less monitoring of action and progress on the SDGs. That isn't to say that nothing is happening. Many cities are doing a great deal under the banner of "sustainability" but haven't made the link to the SDGs. This is a missed opportunity but one that could still be taken. Raising awareness of staff, citizens and other stakeholders at a city level is a tremendous opportunity to help people make the connection between progressive work and a truly global agenda.

Activity 11.2: Monitoring the results of government and others

Where does your curiosity lie? Is it with local government, national government or the United Nations? See what you can find out about the progress they are (or are not) making towards the SDGs. What, if anything, do you feel inclined to do about it?

Measuring corporate progress

The same can be said in organisations, whatever sector they're part of. CEOs like Paul Polman at Unilever and Elon Musk at Tesla are very clear that the work of their businesses is part of the SDG agenda. Around the world, many organisations have signed up to the Global Compact, with many of these making the link to the SDGs.

I truly believe that most organisations make significant contributions to the SDGs through their principle activities. The sad thing is that very few of them do this consciously. Even the obvious contributions are rarely linked to the SDGs and celebrated as such. How many schools have made a link to SDG 4 and told the world how many children they've provided education to; or to SDG 8 and provided data on how their students have found employment and in what disciplines? How many pharmaceutical companies have taken stock of the lives they've saved through their products and linked it to SDG 3? How many engineering companies have devised amazing solutions for sustainable cities, better water supplies, cleaner energy or contributions to a better climate? Do they make the associated connections with the SDGs? Imagine the inspiration they would provide to all stakeholders and what this would do for their businesses? The list is endless and in almost every case, there is a significant win-win to be had.

Activity 11.3: Shining a light on corporate contribution

Think about the organisation you are part of or most familiar with. How does it contribute to the SDGs? What measures indicate its progress? Make some kind of assessment and note it below. Now that you have this information, what will you do with it?

More and more measurement raises consciousness and creates alignment

Imagine the power of more and more nations, governments, organisations and communities being aligned and measuring progress to the SDGs. This is something I truly believe will happen progressively between now and the target date of 2030. As it becomes a way of uniting the world in the delivery of the future we all want, it will surely transition beyond 2030 as the "global way". Who knows it may even help the world to focus on a new "middle ground" that will temper the polarities of consumerism on the one hand and radical extremism of any kind on the other.

Measuring progress in the supply chain

The obvious thing for organisations to manage is the direct contribution their products and services make to the SDGs. However, it's important to recognise the immense and perhaps even greater value that can be gained by measuring SDG contribution in the associated supply chain. How sad would it be for a manufacturing company to make an amazing contribution to the goals through the products they produced, only to discover that the materials used in their manufacturing process caused severe environmental degradation or involved child labour in their production.

The second massive impact of measuring supply chain contribution is that it multiplies the gain several fold. Measuring the direct operations of a company only impacts the work of one company. Measuring what happens in the supply chain impacts several operations and, if they take it seriously, there is a further multiplier in their own supply chain – and so on.

Of course, once an organisation begins to take this kind of measurement seriously, they will probably also feel inclined to communicate with their suppliers and customers why this is so. This further spreads the engagement around the SDGs as discussed in the earlier chapter on this topic.

Families and individuals

We can even measure our progress at a family or individual level. For example, the World Wildlife Fund provides a simple but effective carbon-footprint calculator which can be found at: http://footprint.wwf.org.uk/home/calculator_complete. When I first measured my carbon footprint, I was staggered to see that I was making a footprint that was larger than the world could withstand. I've redressed this through a switch to totally renewable energy, an electric bike and using public transport more often. I now appear to have a much smaller and more sustainable footprint and the very act of measuring this has made me realise where I could be even more prudent.

Activity 11.4: My progress on the SDGs

In the space below, consider as many of the SDGs as you wish (you could try them all if you're feeling ambitious) and list the actions and activities you're engaging in that will help to deliver them. Think what measures you may use to test your progress. If you wish, perhaps you could make a note of your own carbon footprint.

Results or success?

As previously mentioned, results are a tangible, rational measure of our progress towards our vision, which in this case is the world portrayed by the SDGs. Success is very different than that. In *Designing the Purposeful Organization*, I describe success as both an outcome and a feeling. At its most basic level, success is something that we feel as individuals. We will feel success according to who we are, what has happened and how that resonates.

A shared sense of success

However, when people share their sense of success as individuals with each other, something very special happens. We become inspired by the success as described and shown by others. Our empathy for their feelings means that we feel some of what they feel. We see this happening when a crowd responds to a sporting event or an audience to music. The response to a perceived success becomes infectious.

In organisations, I encourage people in teams to think about success in the work they do and to share how this makes them feel. I ask them to then share what inspires them about what they've heard from their colleagues. As this conversation progresses, you can sense a new sense of success emerging, not at the personal level but instead at the collective or team level.

I have noticed a similar behaviour on our SDGs Facebook Page. There are certain posts that I and the other editors of the page make that attract lots of interest and likes from readers. Speaking personally, this affects my publishing actions consciously and subconsciously. Readers seem especially to like posts that describe progress towards the goals, particularly those that are associated with innovation, high achievement or personal heroism. We still post stories about the "other side of the coin" that describe things that are going wrong in the world. These are necessary reminders about why this work is so important. But I think that since we started, there has been a shift to stories that show positive progress.

Activity 11.5: Stock-take of success

Make another list of the SDGs that concern you. List as many as you like. Now think of your greatest achievements associated with each one. Make a note about how you feel about these and about how your feelings inspire you to do more.

My personal sense of success

The greatest senses of success I feel relating to my SDG journey are twofold. They are either to do with personal success gains or the gains of others.

For me, I get a real buzz out of changing my habits and measuring the difference this makes. I take great delight when I clock another hundred miles on my electric push bike instead of driving. I confess to feeling just a little smug when I sit back on the train where I would previously have travelled by car. I take great delight when I see my electricity meter running backwards due to the electricity I produce at home (this will stop when my meter is replaced with a digital one). I feel joy when I see the great work of the Open Arms Infant Home in Malawi or the gains made by Kusamala in permaculture. I was also over the moon when Good Energy put the SDGs at the heart of their brand of energy supply. I don't know how much of these last few "successes" were supported or influenced by my own actions. I suspect it was a drop in the ocean. Nevertheless, this doesn't diminish the delight I take in progress I have been part of.

Shared success

My felt sense of success, however, goes to a new level when I hear of the successes shared by others in my network. The most obvious are the successes we share at UNA and on the SDGs Facebook page. It is early days yet, but I have absolute confidence that the more we share and celebrate our SDG journeys, the more we will be encouraged to do even more.

Now that you are also part of this journey, I trust you will share your successes with your family, friends, colleagues and with me and the SDG community in whatever way makes sense to you.

Chapter reflection

Score the following statements out of ten where:

0 = not at all
2 = a little
4 = moderately
6 = mainly
8 = significantly
10 = completely

1 I appreciate the significant difference between results and success.	
2 I shall be monitoring the results of my national and local governments and companies I work with on delivering the SDGs.	
3 I shall use this data to lobby for more progress and to influence the actions of others.	
4 I have determined the most useful measures of my own progress on SDGs.	
5 I have evaluated my personal SDG results and used this to prompt further personal action.	
6 I have given some thought regarding the goals I particularly care about and the feelings of success I sense when progress is made.	
7 I have shared these feelings of success with others.	
8 I have asked other people I know what goals they care about and what success would mean to them.	
9 I share stories of success on the SDGs Facebook page and on social media generally.	
10 When I see stories of success I congratulate those involved and tell others.	

Chapter 12

Playing to the strengths of humanity

There is plenty of evidence that success in any venture is enhanced by playing to strengths. We will discuss this evidence and provide methods to help people identify strengths and understand how they can be applied to the delivery of the SDGs and developed so they add significant value and be put to use. We will share some case studies ranging from some of the sustainability heroes of the moment such as Elon Musk, founder of Tesla; Malala, who made a stand for the education of girls; as well as some emerging examples drawn from the author's personal experience in sharing the goals with young people in schools and universities. For example, a young scuba diver realised his hobby could be transformed into a career to clean up waste plastic from the oceans (SDG 14). A girl whilst still at school was inspired by SDGs 1 and 16. She became determined to use her flare for photography to photograph poverty, especially in war-torn regions to draw attention and prompt action through journalism.

Take a small dose of positive psychology

In the late twentieth century, with the advent of positive psychology and research into workplace performance, it became widespread knowledge, especially amongst Human Resources professionals, that high performance is largely to be achieved by playing to people's strengths.

Authors like Marcus Buckingham and Donald Clifton in *Now Discover Your Strengths* draw on research such as that undertaken by the Gallop organisation which demonstrates that investing in and playing to strengths gives a better return than over-worrying about people's weaknesses. Daniel Coyle in *The Talent Code* also demonstrates that high performers in all walks of life tend to identify their natural talents and practice them until they add significant value. I dedicate a whole chapter to this in *Designing the Purposeful Organization*.

Connecting to something bigger and better than we are

Once again drawing on fractal mathematics, it seems reasonable to assume that the same logic applies to delivering the SDGs. We are all very different people with our unique sets of strengths. We probably work for organisations that equally have their own strengths and abilities. Some of these are obvious. Teachers and others working in schools, colleges and universities contribute as a matter of course to SDG 4 (education). Doctors, nurses and scientists in the pharmaceutical industry make enormous contributions to health (SDG 3) every day of their working lives. Our engineers and technical people similarly may be solving some of the challenges demanded for sustainable cities, clean energy, adequate water and tackling climate change. Sadly, most of these people and the organisations they work for are not aware of the amazing contributions they are making to the biggest commitment we have ever made as a global society. If they were, it would surely enhance their motivation and encourage others to want to make more of a difference.

Every time I run a workshop on the SDGs, I am impressed by the talent and inspiration of participants. Like the young man at the University of Central Lancashire (UCLAN) who had both a talent and passion for scuba diving. He made the connection to SDG 14 (life below water). He became inspired to explore the possibility of a career cleaning up waste plastic from the oceans. Or the young school girl whose talent and passion was photography and became inspired to apply them to SDG 16 (peace and justice) through journalism in war-torn parts of the world.

Activity 12.1: My natural talents and strengths

What natural talents and strengths do you have? Make note of your top six below:

Activity 12.2: The SDGs I care about

What particular aspect of the SDGs do you care most about?

Activity 12.3: Steps I could take

What are the most useful steps you could take to making the most progress towards the SDGs through the application of your natural talents and strengths?

Activity 12.4: SDGs in my workplace or community

Now repeat the same logic for your company or other community you relate to in some way and make your notes about what you could do to play to these "corporate" strengths.

Big change happens when we play at all levels

Imagine what the world would be like if every person, family, community, organisation and even nation was to understand their strengths and play to them in support of the SDGs. If you think national strengths is a bridge too far, think about China and its manufacturing prowess. In recent years, this country has radically improved its performance and quality in the manufacture of solar panels and now exports them across the globe. Only a few years ago they had very little market share of this new industry and now they are the leaders.

Everyone enjoys playing to strengths

The good thing about playing to strengths is everyone enjoys it, at every level. When individuals play to their strengths, it doesn't feel like work. When companies play to their strengths, they do well and become exciting places to work. Nations take great pride when they discover their niche and grow their reputation like Germany has done for auto engineering, Silicon Valley for IT and London for its financial centre. Of course, the reputation for brand excellence moves with time but the key thing is to build on strength and grow our contribution to the world at all levels.

It's important to recognise that the strengths we bring as individuals don't exist or evolve in isolation. We have often cultivated our strengths according to experience, whether it is as individuals during childhood or as communities or organisations because of many years of experience meeting a particular need. In this respect, it is good to understand and honour our heritage, in whatever form it may have taken.

Outstanding personal contribution

I'd like to provide three "case study" examples of this. Two are well-known people I have already mentioned in this book so far and a third is a personal friend with a remarkable story.

Case study 12.1: Elon Musk

Elon Musk is a serial entrepreneur and CEO having had a series of well-known successes, including PayPal, Tesla, Solar Century, SpaceX and others. He is an amazing visionary, driven by a powerful purpose to enable humanity to thrive sustainably on this planet and beyond. This is reflected by his business ventures in electric vehicles, renewable energy, battery technology and affordable space travel – to name the most famous.

His personal strengths combine to make him a formidable force for positive change. In his childhood he was an avid reader. He became interested in computers as a 10-year-old and taught himself computer programming at the age of 12. He created a video-game called Blastar which he sold for $500. He has two BSc degrees in physics and economics from the University of Pennsylvania. He is clearly a talented applied physicist-engineer, computer programmer, serial entrepreneur and a leader. He is tenacious and hard-working, famous for putting in a hundred hours of hard work a week on a regular basis. He is an engaging communicator, stimulating empowering verbal conversation with the highly intelligent scientists, engineers and others he has recruited. He is prolific on email, using this to engage essential stakeholders in his business ventures.

In terms of context, which often determines how our strengths get played out, Elon Musk was born and schooled in South Africa. Possibly prompted by his avid reading and daydreaming, his childhood had more than its share of unhappiness. He was bullied at school to the extent of being kicked down a flight of concrete stairs before being beaten again and, according to some writers, had an uncomfortable relationship with his father (www.timeslive.co.za/sundaytimes/opinion/2015/05/31/Elon-Musk-How-a-bullied-boy-became-a-man-who-can-change-the-world1). It is thought that he may have found his learning and dreaming to be an escape from unhappiness, as might have been his move from South Africa to Canada and ultimately the US.

I do wonder to what extent Elon Musk was a product of his significant talent being subject to the context of his childhood and all that it brought – as well as living at a time when humanity's very survival was similarly threatened by such immense dangers as climate change and nuclear war. Whatever the case, I believe future generations will look back with gratitude at the life and work of this one man.

Case study 12.2: Malala Yousafzai

My second brief case study of this chapter is of Malala Yousafzai. I introduced Malala in Chapter 2 and described her and others like her as powerful forces in striving for gender equality and education for all young people everywhere, irrespective of their gender or religious background. In this chapter, we are considering the strengths of individuals in the context of the SDGs.

Malala was and is a persistent, confident, persuasive and articulate young woman, drawing strength from her father who believed in her and gave her moral and emotional support. At school she was an avid learner, and her education and that of others clearly meant a great deal to her. These strengths, passions and circumstances seem to have combined to make her who she is. Take any of the pieces of the jigsaw away and her impact could easily have been diminished.

Again considering her context, Malala's strengths were put to good use prompted by significant adversity. It doesn't get much more challenging than being shot and almost losing one's life while simply travelling to school with friends.

Case study 12.3: Richard McCann

Richard had the sad misfortune to be the son of one of the victims of the notorious Peter Sutcliffe, otherwise known as the Yorkshire Ripper. It didn't end there. As a child, Richard continued to suffer at the hands of the adults who were there to provide care for him. Even his own father caused him a great deal of suffering, though Richard found ways later in life to forgive him. Despite, and as a result of, his horrific experience, he learnt to tell his story in a manner that frequently inspires others facing adversity whether as individuals or organisations. Incidentally, he also notes the rare encouragement he received for public speaking from one of his teachers. Richard's first book, *Just a Boy*, telling the story of his childhood, is a best seller and Richard is also one of the world's top motivational speakers, travelling extensively, inspiring many and helping many to climb out of adversity. He has probably helped hundreds of thousands of people out of a variety of poverty traps, though he probably hasn't made the connection between his work and SDG 1. I shall of course be sharing this case study with Richard to make sure he does now.

The power of adversity

I could have chosen any of the case studies from the "Reasons for Hope" chapter and I certainly didn't intend in this chapter to focus on adversity. Whilst well aware of his amazing entrepreneurial accomplishments, I had no idea (before writing this chapter) that Elon Musk had been bullied. When I first discovered this, I still didn't dwell on the fact until I posted an article about him on social media. At that point, a good friend who also had to deal with (a different sort of) adversity pointed out just how many people who go on to do amazing things in their lives had to first overcome some significant personal challenge. She and I discussed the cause of this and I speculated that perhaps facing up to adversity honed such personal qualities as tenacity and willpower. Her response was this:

> We end up in situations we would never voluntarily choose to be in. Somehow we get through it, so adversity changes our perception of

what's possible. That moment of "nothing can ever be as bad as that so I might as well push it as far as I can go." I'm sure there's a psychological term for that somewhere ☺ Keep posting your thoughts and research. I'll be keeping an eye out for your book!

There is nothing to suggest that adversity always played a significant part in all the case studies in "Reasons for Hope" and I'm sure great things have been achieved by people who have led a more ordinary life (if there is such a thing).

The power of context

The thing I am absolutely certain of, though, is this: life, including that of humans with our amazing talents and strengths, does not exist in a vacuum. We exist in a complex context and, just like the stem cells in Bruce Lipton's research, this context determines our sense of purpose. This combines with our talents and strengths and becomes the driver for what we do and what we become, in exactly the same way as Lipton's stem cells became tissue for hearts, lungs or other organs.

This is why I firmly believe that the way to make sure we deliver on the SDGs is not to lecture people with "you should do this or that". It is better to give them the space to consider what is going on in their world and to check out who they are, strengths and all, in the midst of it. Doing this away from the distractions of their busyness allows their sense of purpose to arise. These insights enable us all to encourage each other to make our greatest contributions.

The "energy in the room"

My colleagues and I gained some (albeit limited) evidence to support this recently when we were asked to run a workshop for about a hundred or so professional coaches. We put them into groups of about ten in a large room and asked for a volunteer in each group. This person was then coached by further separate volunteers in turn asking questions along the lines of:

1   What's going on in your world?
2   Who are you?
3   How do you feel?
4   What do you need to do?
5   Is there a commitment you'd like to make right now?

After each question, the skilled coaches were asked to watch for and work with the "energy of the coachee" in the form of excitement, a twinkle in the eye, nervousness and so on. They did this for about ten minutes after each starting question. The effect was startling. It was as if we had "turned up the volume" or "shone a spotlight" on (a) the context of the individual; (b) who they are; and (c) their consequential feelings. After asking question (3) emotions ran high. There were tears in the eyes of some of the coachees. We checked for safety and continued. Everyone involved in the exercise commented on the power of the process, on the power of helping people consider their context, who they are in the midst of it (including their strengths and attributes) and the ensuing feelings. This is, I believe, what drives personal progress.

So, our strengths alone are an important piece of the SDGs jigsaw, but they only come into play when we consider them in our context, as it is now and as we visualise it could be by 2030. Doing this alone is powerful. Doing it with others is mega powerful. Creating all the conditions I originally outlined in DPO (purpose, vision, engagement, structure, character, results, success and talent) in our communities is, I believe, the way to deliver the SDGs.

Activity 12.5: Helping others make connections

Who do you know who makes a significant contribution but hasn't connected what they do to the context of the emerging world as described by the SDGs? How might you help them make that connection? Make your notes below.

"RVDU"

In Chapter 8 of DPO, I describe my philosophy of "talent liberation". It is a statement of belief that actually runs contra to the way talent is managed in many organisations. However, I don't want to go into that aspect right here. But the statement and philosophy is, I believe, crucial to the SDGs.

Organisations reach prime performance when they recognise, value, develop and use the unique talents of all their people in the delivery of their objectives.

By the way, for the purpose of *this* book, the "objectives" are the SDGs and the "organisations" can be any number of humans in any system (such as a family, community, group, team or company).

RVDU has become shorthand amongst my colleagues for "recognise, value, develop and use". These four words represent a series of stages to go through to get the best out of people in any context. So, for the SDGs, starting with ourselves, we need to:

1   Recognise (name and describe) our unique talents and strengths.
2   Understand how these talents and strengths add value in the context of delivering the SDGs.
3   Have a plan to develop our talents and strengths so they can add "maximum value".
4   Make a firm commitment to use our talents and strengths more decisively and more often in support of the SDGs.

This process can then be repeated for others. So we can start to recognise the strengths of others, comment on how these add value in delivering the SDGs, encourage them to develop them further and to put them to use.

Finally, we can put systems in place in our communities and organisations, such as learning and development opportunities and career development plans, to make "RVDU" a way of life.

Conclusion

In summary, making the world a better place in line with the SDGs requires a progressive and concerted effort to provide education to young and old regarding the current state of our world, its strengths and deficiencies and what humanity through its leaders has pledged to do. Against that context, we can help each other to recognise, value, develop and use our personal and collective strengths in the way we feel inclined to do.

Chapter reflection

Score the following statements out of ten where:

0 = not at all
2 = a little
4 = moderately
6 = mainly
8 = significantly
10 = completely

1 I can name and describe my personal strengths and talents. 2 I know which of the SDGs resonate with me personally. 3 I know how my strengths could add value in this context. 4 I have a plan to develop my strengths still further. 5 Considering the above together strengthens my sense of purpose. 6 With this feeling, I have a sense of what I need to do next. 7 I have committed to particular actions that deploy my talents and strengths in support of the SDGs. 8 I have translated the logic of the above five questions to other people, to my family, community, workplace and elsewhere. 9 I know what I need to do next to make good use of the best of who I am. 10 I encourage and affirm positive use of strength, personal and collective, in support of the SDGs wherever I witness it.	

Chapter 13

In closing

This final chapter brings this book to a close or rather it marks the beginning of what is to follow. We recap the journey through the preceding pages and reaffirm some of the key messages. We conclude that there is much work to be done but that none of us needs to be outfaced. We don't need to do it all ourselves or even to take on a "lion's share" unless of course we want to.

Activity 13.1: My biggest learn

When I come to the close of a leadership workshop or even at the end of a day or session, I often ask those involved to name their "biggest learn". This causes people to reflect and work out what was particularly important to them or their organisation. I prefer to do this before summarising because in my summary I will give my headlines which may not be the same as those for participants and may therefore lead them to miss something that could have been important for them. So, in a similar way, before reading this chapter, I encourage you to reflect back on your journey through this book to here. What is it that occupies your attention and prompts further curiosity? What indeed is your "biggest learn"? And what do you need to do after you've finished reading this brief summary chapter? Make your notes in the space below.

It's been quite a journey

When I started writing this book, I had a good plan. It was part of the proposal I submitted to my publisher. But, as with DPO, I didn't really know fully how DPW would play out.

I didn't know that the introduction would begin with the same ideas that I have used several times now to engage people in the workshops I run, by taking you on a trip into the future and asking you to observe the world you'd like to leave for future generations. I didn't know I'd include some of my personal experiences of doing this and the case study of how young people at AIESEC International have run further than I ever could with that particular baton. I didn't expect my introduction to be as long as it turned out.

Nor did I know that my "Reasons for Hope" chapter would embrace all the SDGs with some examples for each that have inspired me personally. I guess I was particularly inspired by the work of Al Gore and, knowing this, I wanted to attempt in some small way to do the same for you, without leaving any SDG stone unturned.

I suppose I knew I would have to say something about my personal journey so we could begin a relationship together. I know that the books I've enjoyed the most were ones where I got to know the author as a friend and fellow traveller. I hope this worked for you and didn't come across as too self-indulgent or as a distraction from the main text.

I always intended to provide a summary of DPO in its own short chapter. For those who have read that book, I hope it served as a reminder of the structure I adopted then and always adopt in one way or another on my consulting assignments. For those of you who haven't seen a copy of DPO, maybe you'll get hold of one if you like my thinking and wish to encourage organisations you know to be more purposeful and inspirational for their stakeholders.

The fractal chapter was a bit of a risk, I guess. For me, fractal thinking is a powerful revelation. It has helped me develop a mindset where I don't get outfaced by the micro or the macro and instead look for patterns that play out at one level and which can be related to another.

The seven chapters that followed, as I'm sure you're aware, are simply a playing out of the DPO principles at a DPW level. In other words they are the results of playing out the well-accepted logic of DPO that has served clients and others my colleagues have engaged with for over thirty years – translating them in a fractal way to create a blueprint for humanity.

I hope you've enjoyed the personal stories, case studies, activities and reflections that have characterised this book. This style is not entirely my own. It is the product of me working with some quite amazing people at Primeast. They are masters of experiential learning and they helped me to make DPO accessible to people with a variety of learning styles. Otherwise DPO and now DPW could have been more about me passing on what I know rather than helping you to discover many answers for yourselves.

So, what now?

Just as it did when I was writing DPO, the creative process of putting this book together has taken me to a new place. Indeed when I wrote DPO, the most exciting chapter was the final one, the one that led me to write this new book.

So, for me, I am left with a new sense of hope despite the current "state of the world". So many people I know are daunted and even frightened by the state of global politics. There are genuine reasons to be concerned. However, shored up by what I know of human nature and the positivity of leading thinkers such as Al Gore, Steven Pinker, Elon Musk, Malala and my other heroes of the moment, I believe the longer term trends will prevail and the apparent turbulence of this decade will pass quickly.

I still believe we will find peace in our world when we invest more in dialogue and collaboration than we do in arms. I believe engineers, innovators and entrepreneurs will find ways to tackle the many challenges we face, including the principal priority presented by climate change. And I believe that the powerhouse of bright, ambitious, compassionate young people of both genders whom we are currently educating will have a much more progressive stance for humanity than we did. It's all good.

For me personally?

Well, for a start, I shall continue to watch, comment on and celebrate the progress we make towards delivering on the SDGs. I intend to spend a greater proportion of my time delivering workshops to raise awareness to the goals in schools, universities, institutions, organisations and anywhere the journey leads. My LinkedIn profile will probably remain the primary source of information about what I'm doing and the thoughts arising from my work in the form of further articles. So feel free to connect. Of course, for those who particularly like what they read in this book, I'll be very happy to deliver keynotes, facilitate workshops or help in other practical ways tailored specifically to what resonates and is most helpful.

I shall also continue to be an enthusiastic "early adopter" and promoter of technologies for clean energy solutions, with climate change mitigation as an end-goal. Look out on social media for news of my solar-powered campervan and my adventures in Europe and beyond fuelled only by sunshine!

Exploring purpose

There also seems to be a bit of a theme for me that began when I first put together the PrimeFocus framework in the 1990s with "purpose" at the top of the triangle. Initially this was simply to do with anchoring and aligning the activities of an organisation, team or individual.

Now I realise it is much, much more than that. As I said earlier, I was significantly inspired by Frederic Laloux in *Reinventing Organizations* when he described how purpose should have a life of its own. I don't know if it's quite what he had in mind but I have come to think that purpose is indeed what life is all about.

In this book we have speculated that the purpose of life is to "thrive in community and celebrate life itself". We have wondered whether this is what "love" is all about.

What if all this speculation is truly getting us closer to a deep and pragmatic view of the meaning of life? With advancements in quantum physics we know that the material world is not what it seems. We know that ultimately all we think of as solid is in fact energy – waves collapsing into form. We know that consciousness is energy and scientists seem to agree that we are more truly energy or spirit than we are material in the form of our bodies.

As we progress more with the mindset that we are not as separate from each other as we seem, my suspicion is we will become more compassionate towards each other and our planet. I am watching this space and I suspect my thoughts and writing will continue the journey of purpose. For those who see any sense in any of this, I trust we will continue to journey together.

Activity 13.2: So, for you, the reader?

I've enquired about your biggest learn and you made your notes about what you might do with such an insight. I'd like to conclude by asking you about where all this is leading? What are the implications for you, your family, your community and those you work with? What will you be doing with the rest of your precious life? What will be your legacy? How will your energy and enthusiasm for life on this planet and beyond be passed to future generations?

Make your notes below before joining me in our final set of ten reflective questions.

Chapter reflection

Score the following statements out of ten where:

0 = not at all
2 = a little
4 = moderately
6 = mainly
8 = significantly
10 = completely

1 Reading this book has increased my awareness of and commitment to the SDGs. 2 I know which SDGs resonate with me personally and why. 3 And I have a plan to support their delivery. 4 I get a sense that I now have more purpose in my life. 5 Before placing this book on my bookshelf, I shall consider who needs to read it next. 6 I know what my biggest learn from reading this book is. 7 And I know what I need to do to turn it into action. 8 I shall give consideration to how I shall work with and alongside the millions of others who are trying to raise the profile of the SDGs. 9 I have connected with the author on social media and look forward to an ongoing journey. 10 I shall drop a note to the author with my feedback on this book.	

Index

Note: Page numbers in italic indicate a figure.

Milton Keynes UK
Ingram Content Group UK Ltd.
UKHW031152141024
449569UK00024B/865